Charles Bray

# A manual of anthropology or science of Man

Charles Bray

**A manual of anthropology or science of Man**

ISBN/EAN: 9783741155604

Manufactured in Europe, USA, Canada, Australia, Japa

Cover: Foto ©Andreas Hilbeck / pixelio.de

Manufactured and distributed by brebook publishing software
(www.brebook.com)

Charles Bray

# A manual of anthropology or science of Man

# A MANUAL

OF

# ANTHROPOLOGY.

"All our hopes now lie in a true understanding and philosophy of man's nature."—H. G. ATKINSON, F.G.S.

"The absolute and unholy barrier set up between psychical and physical nature must be broken down."—DR. HENRY MAUDSLEY.

"It is manifest that nothing can be of consequence to mankind, or any creature, but happiness."—BISHOP BUTLER.

"When religion is called in question because of the extravagancies of theologians being passed off as religion, one disengages and helps religion by showing their utter delusiveness."—MATTHEW ARNOLD.

# A MANUAL

OF

# ANTHROPOLOG

OR

# SCIENCE OF MAN,

BASED ON MODERN RESEARCH.

BY

CHARLES BRAY,

*Author of "The Philosophy of Necessity," "Force and its Mental Correlates," "The Education of the Feelings," &c.*

"If these statements startle, it is because matter has been defined and maligned by philosophers and theologians who were equally unaware that it is, at bottom, essentially mystical and transcendental."—TYNDALL.

LONDON:
LONGMANS, GREEN, READER, AND DYER.
1871.

# PREFACE.

"THE savant," Emerson says, "is often an amateur. His performance is a memoir to the Academy on fish-worms, tadpoles, or spiders' legs; he observes as other academicians observe; he is on stilts at a microscope, and—his memoir finished, and read, and printed—he retreats into his routinary existence, which is quite separate from his scientific."

Cuvier and St. Hilaire disputed over some two hundred pages upon the identity of organs: for instance, whether the forehoof of an ox is exactly the "same organ" with the wing of a bat.

Now we must not for a moment suppose that we have no interest in such memoirs to the Academy, or in such slight differences of opinion as appear to have existed between Cuvier and St. Hilaire; as Mr. Lecky observes, in his "History of European Morals," "in the eyes of both the philanthropist and the philosopher, the greatest of all results to be expected in this, or perhaps any other field, are to be looked for in the study of the relations between our physical and moral natures." Comparative Anatomy is not the least important branch of this study.

Influenced no doubt by such considerations, M. Paul Bert, author of an Essay, "Sur la Vitalité propre des Tissus Animaux," has instituted a series of very ingenious experiments, without, however, as many will think, a sufficient fear of the Society for the Prevention of Cruelty to Animals before his eyes: "Insert," he says, "in the back of a rat the end of its own tail, having first pared it raw with a bistoury; it will heal and take root. As soon as the graft is complete, amputate the tail about one-third of an inch from the old root. The rat's tail will henceforward grow the reverse way and out of the back." And again he says: "If you amputate the paw of a young rat, partially skin it, and introduce it through the skin of another rat's side, it will engraft, take nutriment, grow, and acquire all the ordinary parts of its structure, as if it had remained with its former proprietor." M. Taine, in his recent work, "D' l'Intelligence," has endeavoured to apply these more curious than humane experiments to illustrate the relation between our "physical and moral," or, at least, intellectual nature.

If the researches of the Germans have been less curious than the French, they have certainly been more important. It is now nearly a century since Dr. Francis Joseph Gall, a physician of Vienna, when a school-boy, accidentally made the discovery of a connection between certain definite parts of the brain and certain equally definite mental functions. He devoted his life to this study of Cerebral Physiology: others have devoted their lives to the verification of his most important discoveries, so that the physiology of the brain—"the relation between our physical and moral natures"—is not untrodden and unknown ground. Nevertheless, we find Mr. Lecky, after proclaiming the necessity for cerebral organs, declaring that "of their existence we know nothing." It would have been more correct if he had said *I*—if he had only spoken for himself—although no doubt our leading physiologists appear to be still in the dark as to whether the part the brain has to play is that of a pot of pomatum—its function, according to the barber, being to "percolate through the skull and nourish the roots of the hair"—or to perform the more noble function of the organ of the mind. It is in the latter capacity we recognise it, in its plurality of organs, and we consider the study of the brain and nervous centres, as connected with thought and feeling, quite as important as M. Paul Bert's investigations into "Vitality". —not that we wish to disparage his experiments. Professor Huxley has shown us even the commercial value of Redi's early researches into "spontaneous generation"; and no one, therefore, can say what may come out of M. Bert's transposition of rats' tails from the place where nature intended them, to the middle of the back. The science of the present day, however, would seem to be much more taken up with such speculations, and as to whether our forefathers were covered with hair, had pointed ears and a tail, or whether we began with civilisation, or have ended with it, than with the discoveries of a man who is able to tell us what man *is*, not what he *was* countless ages ago. Gall has been forgotten, and the present generation is much less interested in the head than the tail; nevertheless I am bound to say, notwithstanding the scorn and contempt to which such an assertion will probably subject me, that thirty-six years' close and continued observation have convinced me that the greater part of what has been laid down by Gall, Spurzheim, and Combe is true, and may be verified by any ordinarily well-qualified person who will follow their method of investigation. Of the additions which Mr. H. G. Atkinson has made to their discoveries (for which see "Man's Nature and Development"), as they have been obtained by another method, I have had less opportunity for observation. I can only say I should think them highly probable, as they seem very necessary parts of our mental system. There may be a great deal more to know, and what we do know may be more perfectly known; but upon the facts already discovered has been based a Mental Science, more clear and practical than any with which the world has yet been acquainted, and which is admitted to be "the only psychological system that as yet counts any considerable number of adherents."

All our faculties both of mind and body have been first tried in the lower animals; they have been transmitted to us through every form of organisation; they have appeared singly, separately, or together, until they have taken their last and most perfect development in man. The study of the animal world, therefore, both in its physiology and its psychology, is most important, and it has been too long neglected. We have accepted the love and service of our fellow-creatures of the brute creation, and then meanly denied our relationship. It is true the Common Spelling-book declares the Pig to be man's best friend, because he can eat him every bit, from the tip of his nose to the end of his curly tail; but even in this rather low view of friendship—friendship being *confined* generally to eating with, not of, your friend—it is a question whether man has performed his part. We have ignored all mental affinity with our humble friends, and have suffered, as we deserved to do, from our ignorance of their natures and capacities. We have been studying angels, not animals. In fact,

> In pride, in reasoning pride our error lies;
> All quit their sphere and rush into the skies.
> Aspiring to be Gods if angels fell;
> Aspiring to be angels, men rebel.

In our self-complacency we have carried our noses so high that we have overrun the scent, and must try back. We are gods here on earth; let us not therefore from ignorance fail in our duties to the creatures below us. At present even the languages they speak are to us unknown tongues. We are God's creatures; they are ours to sympathise with, to rule, and to make happy. In all that lives we shall find something to admire, something to learn, and often something to love. Dr. Draper, the American physiologist, says: "Nearly all philosophers who have cultivated, in recent times, that branch of knowledge (Metaphysics) have viewed with apprehension the rapid advances of Physiology, from seeing that it would attempt the final solution of problems which have exercised the ingenuity of the last twenty centuries. In this they are not mistaken. Certainly it is desirable that some new method should be introduced, which may give point and precision to whatever metaphysical truths exist, and enable us to distinguish, separate and dismiss what are only vain and empty speculations."

Only that knowledge which admits of demonstration will endure and advance. Psychology therefore must be brought within the domain of law, if that also, like the other sciences, is to make any progress. This progress, however, must be slow, as the deepest feelings of our nature, and even common sense, have been enlisted and set in array against its discoveries. If it took a century and a half to reconcile mankind to the Copernican Astronomy, it must be long before it will recognise and accept an ethics founded on the Reign of Law, instead of the present one based on chance, called Free-will. We must be prepared for, and tolerant of, some little diversity of opinion on this subject for some time to come, since even the great Lord Bacon said on this question of the earth's motion, "It is the absurdity of these opinions that has driven men to the diurnal motion of the earth; which I am convinced is most false" (De Augmentis Scientiarum, B. III); and Alexander von Humboldt in his early youth —and the majority of the world are still *very young*—wrote a learned essay to refute the notion that the pyramids of Egypt were the productions of nature. Still we may hope that ultimately the "new method" will establish the order of the moral world on a basis as fixed and universal as that of the physical. Why we have no mental and moral science at present is because most people think that they have in their religion a sufficient standard of both, and that all the wants of human nature are included in the Bible; but this idea has long been exploded in Physics, and it must be in Ethics before we can make the same progress in Ethics as we have in Physics. Men agree in Science because it admits of demonstration, but we have no recognised Science of Human Nature. We have no first principles on spiritual subjects: consequently on great moral questions that arise we find our most talented and leading men about equally divided on opposite sides. Talent without principles is rather in the way than not, and interferes with the instincts that would otherwise guide us aright, and

> Who fastest walks, but walks astray,
> Is only furthest from the way.
>
> —Prior.

And we have a Babel of discord and a confusion of tongues.

Anthropologists occupy themselves with questions relating to facts "so far off and so long since" that they scarcely admit of demonstration and verification. Perhaps they do not know that the *exact* age of the world has been determined. This was accomplished, according to Venerable Bede, at a council held at Jerusalem, about the year 200 A.D. After a learned discussion, reported *verbatim*, it was finally decided that the world's birthday was Sunday, April 8, at the vernal equinox, and at the fall of the moon (Opera, tom 2, pp. 346, 347. Ed. Basil, 1563.) Here, at least, Anthropologists have *one* valuable and important fact to start from, and this question having been so satisfactorily settled, they may perhaps feel disposed to take the next step onward; so that, in course of time, arriving at the idea that sensibility is connected with the nervous system, and that the brain is the centre of this system, they may begin to feel some little interest in its varied functions and Gall's discoveries.

There are many at the present time who have been obliged to disregard "the tales they have heard from their mothers," and to abandon prevailing opinions, anxiously enquiring what is to take their place, and they are as desirous as the world has ever been to know Whence we came, Why we are here, and Whither we are going. The *Spectator*, a paper of some authority in such matters, in a notice of Mr. Maurice's work on the Conscience, September 26th, 1868, remarked that a large section of the ablest of our young men hold and avow more or less openly, according to their courage and honesty, that the attempt to solve such problems as the moral philosopher deals with "can lead to no results save that of entangling the inquirer in vague speculations, incapable alike of refutation or verification;" and Lord Macaulay also asserted that "the best writer on Morals does not deserve half the gratitude from mankind which is due to a good shoemaker." Nevertheless the present work is another brief and humble attempt, by making use of the light of modern discovery and by putting together and systematising what have hitherto been detached and isolated truths, to answer this question of our Whence, Why, and Whither. I have endeavoured honestly to think out modern facts and discoveries to what appear to be their legitimate conclusions, although it may perhaps take the labours of another generation of workers to test and verify the deductions that have been here made. This task has been done without dread of the consequences, in the firm conviction that we have nothing to fear from truth, and that whatever intuitional aid we may receive from Conscience, "we must still learn what is true in order to do what is right."

The Author has been accused by former critics of being too fond of "inverted commas," that is, of quotation. This accusation is quite true. Having no personal ends to serve, and knowing the value of authority with the general public—for much that is here if given merely as his would probably obtain little acceptance—he has always been glad to support his position by names deservedly better known to the world than his own, and to build up his edifice with bricks that have the "trade mark" of science upon them. It has also been said that he offers "authority" for "proof;" this also is partially true. In a single volume it is impossible in many cases to do more than point to where proof may be found, and those who are really in search of truth, and wish to study the subject, may refer. Should such a truth-seeker meet with apparent contradictions, let him reflect how very imperfect a medium language is to express new ideas, and that we are frequently obliged to use old terms in a new sense, which often begets an apparent contradiction. Men ordinarily believe, like Shakespeare's Shepherd, "that the property of rain is to wet, and fire to burn, and that a great cause of the night is lack of the sun;" they believe also quite as firmly that grass is green, and that there are forms and bodies to which this colour belongs: they never for a moment suppose, that they are conscious only of their own perceptions, and that that consciousness is all they know, or can by possibility know; and yet as long as they continue under this illusion and delusion, and fail to recognise that the greenness is in themselves, and not in the grass, the same apparent contradiction must appear in language as there is in that which speech is supposed to represent.

It has also been said that the doctrines here propounded tend to lower our ideal. This also may be partially true, but our real good will be found in the proper restraint of our unbounded ideal longings within proportions commensurate with the real facts of life. Much of our duty may be better done by the light of a farthing candle than by that of the stars. No doubt much else may be justly said, much unjustly; but

> To the long-necked geese of the world that are ever hissing dispraise,
> Because their natures are little,
>
> —Tennyson,

The author has nothing to reply.

The Spectrum Analysis is affording proof to many problems in physics, and Prof. Tait tells us that the doctrine of the Conservation of Energy is now thoroughly accepted by scientific men, and that it has already revolutionised the greater part of physics. Scientific men, however, have confined this discovery to physics, and there has been little or no attempt hitherto to carry it beyond. The Bishop of Gloucester and Bristol, however, says "that we owe much to Dr. Carpenter for making it perfectly clear that though there is a certain amount of correlation between vital and physical forces, yet that the *differentia* between them is distinct and well defined, and that it is to be sought for in the nature of the material substratum through which they work, whether that be inorganic matter or organised structure. Such generalisations are helpful and suggestive; we owe much to them, and in the future we may owe still more." I have endeavoured to carry this generalisation a step further, to the correlation between *mental* and physical forces, and all the novelty I lay claim to in this work is the application of the Conservation, Transmutation, and Dissipation of Energy to Mind, Morals, and Religion. This doctrine of the Persistence of Force or Conservation of Energy furnishes the proof to much that has been known from the earliest ages, but which has been known as mere speculation. Prof. P. G. Tait tells us that it is "simply preposterous to suppose that we shall ever be able to understand scientifically the source of consciousness and volition." Nevertheless, I have endeavoured to show the conditions under which physical force or *automatic mind* again resumes its consciousness; how the Persistence of Force, and Philosophical Necessity or Law in Mind, are the same, and how, therefore, our Ethical systems may and must be brought into harmony with this now known fact; and I have endeavoured also to show the Unity of Force, and that all Power is Will Power, conscious or automatic, or, as Mr. W. R. Grove has put it, "Causation is the Will, Creation the act of God." Physics and Metaphysics—Physiology and Psychology thus become united, and the study of man passes from the uncertain light of mere opinion to the region of science.

I wish some one with a learned handle to his name, or with a better claim to the world's notice, had taken up this subject; but the fact is that literary and scientific men are either too much occupied in their special department, or do not like to face the *odium theologicum* and all the other odiums which attend speaking out on subjects which in the present day belong more to feeling than to reason. It is only one who has no position and reputation to lose who dares venture upon this ground. Dr. Tyndall, in the Preface to his Volume of Essays just issued, says: "Most of the Essays are of a purely scientific character, and from those which are not I have endeavoured, without veiling my convictions, to exclude every word that could cause needless irritation." I am afraid I have not been equally cautious. I bow to science, but science is only valuable as it helps to solve the problems here propounded of the Whence, Why, and Whither. Faraday, however, says ("Life and Letters"): "I do not think it at all necessary to tie the study of the natural sciences and religion together, and in my intercourse with my fellow-creatures that which is religious and that which is philosophical have ever been two distinct things"; and, if we are to follow Faraday's example, that which is religious and that which is philosophical must ever remain two distinct things. Faraday was a "Sandemanian," and if we are precluded from the exercise of reason in Religion, and are not "to tie the study of the natural sciences and religion together," we can have no hope of finding God in Nature, or of discriminating between the grossest superstition of the age and country in which we are born and the Religion of the Universe. Sir John Lubbock bears his testimony "that without science true religion is impossible." Faraday, however, was not without precedent. "The most enlightened theologians of the Catholic Church—Pascal, Malebranche, Bossuet, and Fenelon, received what they called Catholic doctrines, and mysterious dogmas to which no principles of reason could be applied. Some even said that the more the mysteries shocked the reason and the conscience, the more devoutly they were to be believed." "In all superstition," says Lord Bacon, "wise men follow fools; and arguments are fitted to practice in a reversed order." Again, the Duke of Argyll (*Contemporary Review*, May, 1871): "I do not know that the discoveries of modern science, great as they have been, and much as they are vaunted, have contributed anything towards the solution of the final problems of all human speculation. These, in so far as mere speculation is capable of dealing with them, seem to remain very much where the great intellects of the ancient world found them and left them." Surely Science, with those who, unlike Faraday, think it right to use it, has taken many of these problems out of the field of mere speculation. It has tested more than one Revelation, and shown that the sun no more goes round the earth in Ethics than in Physics, however in the one case, as in the other, appearances may deceive us.

COVENTRY, *October*, 1871.

# CONTENTS.

## CHAPTER I.

### IN THE BEGINNING.



## CHAPTER II.

### MAN.

## CHAPTER IV.

### PHYSICS AND METAPHYSICS.



## CHAPTER V.

### RELIGION.

## CHAPTER VI.

### SOCIOLOGY.

## CHAPTER VII.

### SUMMARY AND CONCLUSION.

# A MANUAL

OF

# ANTHROPOLOGY,

ETC.

---

## CHAPTER I.

### IN THE BEGINNING.

It is very difficult to fix dates in Geology; before or after certain occurrences is all that can be spoken of confidently with respect to time.

If we admit, therefore, that life may have existed upon this earth from one hundred millions to five hundred millions of years, this latitude, great as it seems, is perhaps no greater than the present state of our knowledge requires; and it is fortunate that a hundred million years, more or less, make very little difference to our comfort at present. Authority is rather in favour of the smaller estimate.

Our earth and planetary system, and indeed the whole Cosmos, are supposed to have existed originally as nebulous matter—that is, a sort of "fiery mist" like the tails of comets. Our system has gradually settled down into its present form, evolving in the process an enormous quantity of heat, so that our earth must at one time have been molten at the surface, with a fiery atmosphere like the sun, and probably being to its satellite the moon what the sun is to us. The moon may be a dead and used-up world—such as the earth is becoming, and may probably become, before it falls into the sun, towards which all planetary bodies are gravitating. The sun may then take a habitable form, and revolve as a great planet round some more distant centre, and it and other stars in our "Milky Way" may form Planetary Systems on a larger scale, or may, by the great heat consequent on the change of their present "mode of motion," be again melted into nebulous fiery mist, and begin life again, as our earth is supposed originally to have begun. The earth has been very gradually cooling down, and is still conjectured to be in a state of internal fluidity. According to Sir Charles Lyell, the heat increases as we leave the surface for the interior one degree for every 65 feet. This degree of heat would boil water at a depth of two miles, and melt iron at a distance of thirty-four miles. Mr. Hull, in a paper read before the Royal Society (Jan., 1870), stated that at Rose Bridge Colliery, near Wigan, the deepest mine in Britain, it appears that at a depth of 808 yards the temperature of the coal is 93½ degrees. This gives an increase of temperature of one degree for every 55 feet, and will render the supply of coal, which geologists say exists in such large quantities beneath the new red sandstone, rather difficult to reach, until a considerable farther cooling of the earth has taken place. At the meeting of the Newcastle-upon-Tyne Chemical Society (1871) it was however stated that salt-mines at nearly the same depth as the Monkwearmouth Colliery are at a much lower, in fact, a comfortable working temperature; and the conclusion drawn from this was that the heat is due to chemical action going on within the coal itself. As respects this internal fluidity a considerable influence may be expected to be exercised upon the peculiar "mode of motion" which we call heat by the great pressure upon it from above.

But whatever may be the present peculiar state of the earth's interior, it seems to have existed in a very unquiet and uncomfortable condition for a very considerable period, and great commotions have taken place on its surface. Mountains, much higher than any at present existing, have been raised, continents and islands everywhere have emerged to be again submerged; and to come as near home as possible, it is Sir Charles Lyell's opinion that at one stage of the glacial period the mountains of Wales were much higher than at present;—at another stage, 2,300 feet lower; at one stage of the same period Scotland was 2,000 feet below its present level, and other parts of Britain 1,300 feet; then Great Britain and Ireland consisted of a few groups of small islands formed by the mountains of Wales, Cumberland, the Scottish highlands, Munster and Connaught, and probably of a larger island formed by the portion of England south of the Severn and the Thames. Thus, through the combined influences of Catastrophism, Uniformitarianism, and Evolutionism,—representing the three great schools of geological theory,—the earth has settled down into the comparatively quiet state in which we now find it, atmospheric and aqueous agencies having given it the varied strata so familiar to us. The crust of the earth is supposed to be about 60 miles thick, which bears the same proportion as one inch to a globe eleven feet in diameter. We need not be surprised therefore if some still further upheaval and lowering of continents should take place, and in fact it is found that the Western Coast of South America is steadily rising, and in North Europe the lands are in some places rising and others falling. Much of Scotland, Scandinavia, and of the Baltic shores have greatly risen in these latter times, and *Cosmos* asserts that it has recently been demonstrated by a reference to authentic documents that Guernsey and Jersey have sunk more than fifteen yards during the last five centuries. At present two-thirds of the earth's surface are covered with salt water.

By the aid of the spectroscope it has been demonstrated that planets, sun, and stars are all made of one material,—that the matter of the sun is the same as the meteoric stones that occasionally fall heavy and cold to our earth. In the star Aldeboren nine like elements have been detected, and other stars have the same elements as our earth and sun, or very nearly—which induced a very venerable archdeacon, ever on the look-out to illustrate the relation between religion and science, to observe that this showed that they had the same Creator, *or very nearly*. The sun, being a much larger body than its satellites, still appears to exist in the incandescent molten state from which our earth has so gradually emerged, and storms of fiery gases still sweep over its face, causing considerable physical and mental commotion on earth. The great frost at Brussels, February, 1870, in which the centegrade thermometer went suddenly 10 deg. below zero, and all the lakes, canals, and ponds were covered with ice, was observed to be coincident with two sun spots, visible to the naked eye, and measuring in diameter twice or thrice the diameter of the earth. In 1866 a small star in the Northern Crown was seen to flare up into a most abnormal splendour, and as suddenly to collapse into its original state. This was supposed to be owing to some disturbance in its fiery atmosphere; probably from an unusual conflagration of hydrogen gas.

But what has become of the heat in this process of the earth's cooling down? Professor Tyndall tells us that "ages ago the elementary constituents of our rocks clashed together and produced the motion of heat, which was taken up by the ether, and carried away through stellar space, and was thus lost for ever as far as we are concerned." This, however, is true only in a very partial and limited sense. Heat is force or energy, but is known to us only as a "mode of motion." "The atoms of matter," says Professor

Tyndall, "are suddenly urged together, by their own perfect elasticity they recoil; and thus is set up the molecular oscillation which announces itself to the nerves as heat." On the principle of the Persistence of Force, or Conservation of Energy, a comparatively late discovery, and which Faraday pronounced to be "the highest law in physical science which our faculties permit us to perceive," no force is destructible: it merely changes its form, or mode of motion, and all the phenomena in the universe consist but in these changes of form or transformation of energy. Thus heat, light, electricity, chemical affinity, life, mind, are forces known to us only in their modes of motion, which Mr. Grove has shown, at least as far as the physical forces are concerned, to be readily transformable one into the other; and the peculiar mode of motion, which we call heat, has not been lost to us, but has merely changed one mode of motion for another, and shows itself in all the varied phenomena of inorganic and organic bodies around us.

Heat appears to be the most diffuse or least concentrated form of force; mind is the most condensed, and is most dependent upon all the others.

The heat carried "away through stellar space" is more than compensated to us by that which we receive direct from the sun. It is the sun which lifts the seas from their beds, with a power greater than that exerted by all animated creation, and depositing them on the tops of mountains, produces that circulatory system upon which the vitality of the earth depends. It is the sun, divorcing the carbon from the oxygen in the plant, and on the restoration of this force by their re-union in the animal body, which produces all the power, bodily and mental, that animals display. Plants are fed on air, not on earth, their substance being formed principally of the carbon they abstract from the atmosphere; all they get from the earth is the small quantity of ash left on being burned, *i.e.*, on being again converted into air. As

Professor Huxley says, "Plants are the accumulators of the power which animals distribute and dispense."

Philosophers recognise at present only two sources of power or energy—Sun-force and Earth-force; but probably there is only one, and gravitation and the special liking of atoms for each other (chemical affinities) are dependent on the sun,—*all* forces being *equally* correlates of each other.

The same force transmitted through the same impalpable *ether*, or rather the same force and that *unknown spiritual essence* of which it *is* the force, may equally hold the planets to their appointed places and curve their far-off wanderings, and create those undulatory and other modes of motion in our atmosphere on which light and heat depend. The light of day and the blue sky are owing to the comparatively solid state of our atmosphere. The deep blue, Professor Tyndall tells us, is produced by particles suspended in the air, which are not only invisible to the naked eye, but irreducible by the highest powers of the microscope, the light which they scatter and their power of producing colour being indubitable evidence of their existence. Were it not for this, we should look into "the darkness of stellar space" instead of the bright blue sky. These "suspended particles" are living "germs," always ready to fructify under favourable conditions. From experiments, aided by the microscope, Dr. Angus Smith, of Manchester, and Mr. J. B. Dancer found that in the quantity of the air of Manchester that a man would breathe in ten hours there were 37½ millions of these spores or germs of organic matter. This would make it very difficult to prove the fact of spontaneous generation. Professor Tyndall says, "Apart from their other effects, the mere mechanical irritation produced by the deposition of these particles in tender lungs must go for something. They may be entirely withheld by a cotton wool respirator."

It might be thought difficult to prove that these dust particles were living germs; so also it might be considered to be at least difficult to prove their existence at all, as they are invisible, and beyond the powers of the microscope; but taken *en masse* their existence is proved by the light which they scatter, and their organic character is revealed through the spectroscope, by the light which they give when burning.

Light and Heat depend quite as much upon the atmosphere that surrounds a planet as upon the sun's energy, and we may suppose that the atmosphere may differ from ours quite sufficiently to compensate the planets that appear to be left out in the cold for their greatly increased distance from the sun. The mode of motion or quantity of heat may therefore be the same in each planet. Saturn is of about the same substance and weight as cork; this, perhaps, may be traceable to the conditions under which it cooled, but this levity would probably necessitate a difference in the organic form of the creatures upon it as great as exists here between the inhabitants of sea and land. In the sea the animals have no weight, being of the same specific gravity as the element in which they live; which must be very convenient. Recent solar researches would indicate that the solar atmosphere consists of clouds of iron, copper, zinc, and other such elements, in a gaseous state; each element having its own distinct range, and the clouds being consequently in layers.

Astronomers now rejoice in the knowledge of 88 planets instead of 7, and how many more there are, time and our telescopes can alone determine. Mars is supposed most nearly to resemble our earth. Its year is almost twice as long as ours; its density less than three-fourths that of the earth, and the force of gravity therefore much less than half.* Friction thus being proportionally less, the same amount of force would cause carriages and everything to move faster. As much less force would be required for locomotion, it might be used in brain work, that is in thinking and feeling,

* "Other Worlds than Ours," by R. A. Proctor.

and its inhabitants would be much more spiritual. Velocity of thought as well as of motion, and intensity of feeling, would probably be the distinguishing features of life in Mars.

When the earth was without life and mind it was the same as if it had no existence; for a world that is not perceived is the same as no world at all; and we shall so regard it, for, as we shall find, the objects of knowledge are ideas, not things. Its real creation was in the perception of it by the creatures upon it, and then only so much of it as was perceived. Mind we find only in connection with nervous matter, and it will very much simplify and shorten our course if we confine ourselves principally to the gradual production and increase of that nervous matter, and omit the preparation for it as illustrated in geology and its countless ages and guesses.

Professor Huxley considers that as worlds and systems have grown out of a comparative chaos of nebulous matter, so all animal life has been evolved from a shapeless mass of what he calls protoplasm. To such a mode of creation we have only to allow plenty of time; and instead of one day, man must have been one hundred millions of years in creating at least. "Beast and fowl, reptile and fish, mollusk, worm, and polype, are all composed of structural units of the same character, namely, masses of protoplasm with a nucleus—lower down still are particles without a nucleus"; and Professor Huxley says "it is a fair question whether the protoplasm of these simplest forms of life, which people an immense extent of the bottom of the sea, would not outweigh that of all the higher living beings which inhabit the land put together."* The earliest inhabitants of the earth appear to have lived in shells, and to have carried their houses about with them. A cubic inch of chalk contains

* "On the Physical Basis of Life."

the bones, that is, the shells of 10 millions of these doubtless highly-organised living creatures. M. Ehrenberg ascertained that at least 57 species are still existent in different parts of the earth, and if we take into consideration their numbers, and even their capability of enjoyment, they are probably of greater importance than higher animals. Still, in calculating "the greatest happiness of the greatest number," on utilitarian principles, it will be scarcely fair to allow each to count as one.

Professor Huxley tells us "the microscopic fungus multiplies into countless millions in the body of a single fly." If a drop of water can contain 500 millions of living animalcules, each manifesting very decided volitions, and therefore endowed with mind and muscle; if the bodily and mental constitution even of man can be transmitted through an invisible "germ," it is evident that our senses can tell us as little as to what constitutes organisation, and the degree of its perfection, or of its spontaneous production, as they can of the invisible floating particles in the atmosphere. The first organ is a stomach, as the assimilation of food is the first condition of animal life. The first animal is all stomach, and it digests as well on the outside as the inside; in the lower organisms all parts being competent to perform all functions. This organ has continued in almost equal importance up to the present day, for not only are there men who are all stomach, that organ governing if not performing all other functions, but it is the foundation not only of all physical strength, but of all moral order. The source of the development of all our faculties, and cause therefore of all our happiness, is exercise, that is, work; and if a man will not work, nature pinches him in the stomach, and he is soon obliged to fall in and do his duty in the ranks with his fellows. Hunger lies at the base of all moral order. It may be a very low motive, but it is the most imperative and efficacious of all.

But if digestion was a primary condition of animal life, so

was the continuation of the species. "To draw nutrition, propagate, and rot," were therefore the first functions differentiated, and some among creatures considered higher in the scale appear to have been quite willing to stop there.

> Eat, drink, and love, what will the rest avail us?
> So said the royal sage Sardinapalus!

Some of these lower creatures seem to have had rather large families. Ehrenberg estimated that a single married couple of *Hydatina* or Rotifers, or animalcules, are capable of multiplying to 17 millions in 24 days* ; an oyster, a highly-organised and very loving animal, has a small family of about 8 millions annually; and higher up in the scale many fishes eject a million spawn in a single year; and if *justice*, in the human sense, were done to each individual, whether of animalcule or fish, they would soon have the world to themselves.† Higher still: "A naturalist one day took a butterfly that comes from a hairy caterpillar, and obtained from it a brood of 350 eggs, which all hatched. He kept 80 eggs, and brought up the caterpillars. All performed their changes and became perfect insects, except five, which died in the act of changing their skins. Among all these butterflies there were 15 females coming from the 80 eggs; the brood of 350

* "An animalcule, visible only under a high magnifying power, is calculated to generate 170 billions in four days," and "these enormous powers of propagation are accompanied by a minuteness so extreme, that of some species one drop of water would contain as many individuals as there are human beings on the earth." * * This prolific power is not confined to the animal world: "Before a cocoa-nut tree has ripened its first cluster of nuts, the descendants of a wheat plant, supposing them all to survive and multiply, will have become numerous enough to occupy the whole surface of the earth."—Herbert Spencer, "Principles of Biology," vol. 2, pp. 423, 465, 431.

† Man is the only animal, however, who *claims* to have as large a family as he pleases, without reference to whether he can keep them or not; witness, for instance, the outcry about the separation of man and wife in our workhouses.

would then in proportion have furnished 65 females. Supposing that they had been as fruitful as their mother, we come in the third generation to a sum of 22,750 caterpillars, among whom there would have been at least 4,265 females, and they would have produced in their turn 1,492,750 grubs. These numbers are sufficiently oppressive; but certainly viviparous flies will show as much larger ones. They produce as many as 22,000 at a single birth. For these, supposing the number of females equal to that of males, we shall have at the third step not a miserable million of individuals, but a splendid posterity of two thousand decillions of beings, all living, flying, and eating at the universal board.

"The *musca carnaria*, or blow-fly, in 24 hours becomes 200 times as heavy as at first.

"The raisers of silkworms have proved that the caterpillars hatched from one ounce of eggs eat 1,609 mulberry leaves before reaching the perfect state. This enormous consumption of eatables is necessary to the larvæ, as they must lay in a store of sustenance to be able to bear the long fast they undergo in the state of chrysalis or nympha. Their stomach changes with their nature. The greedy silkworm that consumed a mass of green leaves 60,000 times its own weight, when become a moth only takes a little honey by way of nourishment. The digestive organs of the first had very great power; those of the second are reduced to the thickness of a thread.

"Poor earwig! unfortunate worm! why have they not the impertinence of the Coach-horse, or the credit of the Lady-bird? Certainly it is not from man that they have learnt the lesson of humility and modesty. From the time he put in an appearance here below, he made the earth that he inhabits the centre of the universe, and himself the point of attraction to everything in this world. Believe him, and stones, plants, animals, everything, have only been created with a view to his employment of them. This theory still numbers many votaries, and would be pleasing to me also did not some facts stand in singular opposition to it." *

"Every animal has at first the form of an egg, and every animal and every organic part, in reaching its adult state, passes through conditions common to other animals, and other adult parts. Every animal, however apparently different—men, horses, birds, reptiles, fishes, snails, slugs, oysters, corals, and sponges—is organised upon one or other of the five, or more, plans which zoologists have adopted for their classification. The ape, the rat, the horse, the dog; and then the bird, the crocodile, the turtle, the frog, and the fish are admitted into the same sub-kingdom of the *Vertebrata.*

"All the great classes of animals, beasts of the field, fowls of the air, creeping things, and things which dwell in the waters, flourished upon the globe long ages before the chalk was deposited. 'Very few, however, if any, of these ancient forms of animal life were identical with those which now live. Certainly not one of the higher animals was of the same species as any of those now in existence. * * * The population of the earth's surface was at first very similar in all parts, and only from the middle of the Tertiary epoch onwards began to show a distinct distribution in zones. * * The constitution of the original population, as well as the numerical proportions of its members, indicates a warmer and, on the whole, somewhat tropical climate, which remained tolerably equable throughout the year. The subsequent distribution of living things in zones is the result of a gradual lowering of the general temperature, which first began to be felt at the poles. * * There are two hundred known orders of plants; of these not one is certainly known to exist exclusively in the fossil state. The whole lapse of geological time has yet yielded not a single new ordinal type of vegetable structure.

* "The Population of an Old Pear-Tree," by E. Van Bruyssel, pp. 174, 191.

"The positive change in passing from the recent to the ancient animal world is greater, but still singularly small. No fossil animal is so distinct from those now living as to require to be arranged even in a separate class from those which contain existing forms. It is only when we come to the orders, which may be roughly estimated at about 130, that we meet with fossil animals so distinct from those now living as to require order for themselves; and these do not amount, on the most liberal estimate, to more than about ten per cent. of the whole." *

The first form of life is the simple cell and cellular mass; and passing on to these interesting creatures with the large families, who try to make up in quantity what they want in quality, as the conditions for the higher life are gradually prepared, we go on through fishes, reptiles, birds, and beasts, up to man, who himself passes through all these stages in his earliest or fœtal state: the mark of their origin being left on the bodily form of too many of the superior race in after life. In one man you distinctly trace the fish, in another the bird, or other animal, and some are reptiles all over. In the geologic strata, in the graves of the earliest races, a clear progress is thus traced throughout from humble to superior types. As suitable conditions arose—for the land had to be *elevated* before its inhabitants could be—life in some form pressed in, and as these conditions were again withdrawn, so were the particular forms then living. In these forms, however, one fundamental plan seems to have prevailed throughout, the varieties being merely modifications to suit the new conditions. We find throughout the entire organic world an indissoluble unity, which can only be explained by similarity of origin.

"Carbonic acid, water, and ammonia," says Professor Huxley, "disappear, and in their place, under the influence of pre-existing living protoplasm, an equivalent weight of

* Huxley. "Persistent Type of Life."

the matter of life makes its appearance," and he also says "a unity of power or faculty,-a unity of form, and a unity of substantial composition, pervades the whole living world. The difference between the powers of the lowest plant or animal and those of the highest is one of degree, not of kind. * * Plants can manufacture fresh protoplasm out of mineral compounds, whereas animals are obliged to procure it ready made, and hence, in the long run, depend upon plants. Upon what this difference depends is unknown." *

But it is with Sensibility, Mind, and Consciousness, that I propose to deal; and sensibility or feeling is everywhere connected with the nervous system, and with complexity of function is always allied complexity of structure. It is difficult to say where sensibility is first added to organisation, or to trace a nervous system in those minute creatures which show evidence of volition, and therefore of power of mind. Mental energy, however, increases as we rise in the scale of creation, and always with it the enlargement and complexity of the nervous system. "The brain, which bears an average proportion to the spinal cord of not more than two to one, comes first,—it is the brain of the fish; that which bears to the spinal cord an average proportion of two-and-a-half to one succeeded it,—it is the brain of the reptile; then comes the brain averaging as three to one,—it is that of the bird. Next in sucession came the brain that averages as four to one,—it is that of the animal; and last of all there appeared a brain that averages as twenty-three to one,—reasoning, calculating Man had come upon the scene." † This is the order in which the four great orders of vertebrate animals appear in the order of time. But there is a world of acute intelligence outside these orders: or, rather, I ought to say thousands of worlds, for in each creature is created a world peculiar to itself, according as the so-called qualities and properties of

* "On the Physical Basis of Life."

† "Footprints of the Creator," by Hugh Miller.

matter or the forces without it play upon its more or less varied nervous system. Every minute space has its living creature whose wants are adapted to its peculiar locality and condition, and whose mind or intellect is sufficient to guide it towards the full fruition of those wants. Man's intellect is very inferior in certain departments to either that of the Insect or the Bird. What is a man's sight, for instance, to that of a fly, with its hundreds of microscopic eyes;* or his hearing to that of a gnat's, the music of whose wings is owing to their making ten or fifteen thousand strokes per second? Although each stroke requires a separate nervous action, probably the motion is automatic, or not attended by consciousness; still, if time is measured by motion, as with us, or even by ideas, the life of a gnat may be much longer than the life of a man. In smell can man equal the dog? or the pigeon, or even a poor *blind* seal, in finding his way about without finger posts? Not even in the same species does the same amount of outer force or properties of matter excite exactly the same kind of ideas and feeling: how much less, then, in such varied organisms. "It is certain," says Darwin, "that there may be extraordinary mental activity with an extremely small

* "Not only has it 4,000 eyes, instead of two; three sets of brain or nerve-centres, instead of one; 1,000 hairs and two claws instead of toes on its foot; it has also wings, which we have not; three pairs of legs instead of one; a mouth which would bewilder any dentist; and a proboscis as far beyond that of an elephant in complexity of structure as a railway engine is beyond a wheelbarrow."—Dr. Tristram, *Sunday Magazine.*

"Insects are commonly cited to express ideas of *smallness.* But to innumerable creatures they are what whales and elephants are to ourselves. The animal that holds the middle place in the scale of size, reckoning upwards from the *Monas crepusculum*, the minutest to which our microscopes have reached, is the common house-fly. That is, there are as many degrees of size between the house-fly and the Monas, reckoning *downwards*, as, reckoning upwards, there are between the house-fly and the whale."—"Life, its Nature, Varieties, and Phenomena," p. 81, by Leo H. Grindon.

absolute mass of nervous matter: thus the wonderfully diversified instincts, mental powers, and affections of ants are generally known, yet their cerebral ganglia are not so large as the quarter of a small pin's head. Under this latter point of view, the brain of an ant is one of the most marvellous atoms of matter in the world, perhaps more marvellous than the brain of man." * * * "Ants communicate information to each other, and several unite for the same work, or games of play. They recognise their fellow ants after months of absence. They build great edifices, keep them clean, close the doors in the evening, and post sentries. They make roads, and even tunnels under rivers. They collect food for the community, and when an object too large for entrance is brought to the nest they enlarge the door, and afterwards build it up again. They go out to battle in regular bands, and freely sacrifice their lives for the common weal. They emigrate in accordance with a preconcerted plan. They capture slaves. They keep Aphides as milch-cows."* But it is the power of feeling in man, proportioned to his volume of brain, in which man's greater superiority consists. Insects do many things in their worlds that he could never accomplish. It is true their field of view is very limited, as they perceive only so much of the world as acts upon their limited faculties; or rather, perhaps, I ought to say, *the same force* acting upon their limited nervous systems creates but a very small world for them, but its wonderful perfection is in proportion to its limitation.

A network of nerve of varying powers or functions is thus gradually spread over the whole world; consciousness and feeling varying with varied structures, but all being of the same substance and chemical composition and obeying the same laws—the same molecular action of brain or nerve in all cases preceding consciousness. Man equally obeys the laws of his being with the minutest creatures around him, and is

* "Descent of Man," vol. 1, pp. 145, 187.

equally a part of one great nervous system, where "each manifestation of force can be interpreted only as the effect of some antecedent force." The difference between the lowest and the highest animal is one of degree, not of kind; and the essential differences that man creates have no real existence, but are entirely the creations of his own mind, that is, are purely subjective. "Matter and law have devoured spirit and spontaneity. And as sure as every future grows out of every past and present, so will the physiology of the future gradually extend the realm of matter and law until it is co-extensive with knowledge, with feeling, and with action. * * * For after all, what do we know of this terrible 'matter,' * except as a

* On this subject we would refer the more curious reader to "The Mystery of Being; or are Ultimate Atoms Inhabited Worlds?" By Nicholas Odgers. Redwith and London. 1868. 8vo. Also to "Spectrum Analysis." Six Lectures delivered in 1868 before the Society of Apothecaries, London. By H. E. Roscoe, F.R.S. Macmillan and Co. Mr. Roscoe shows that by the Spectroscope we are able to detect the eighteen millionth part of a grain of sodium in a room; to determine the constituent elements of stars so remote that light, which travels some 180,000 miles a second, requires more than 100 years to reach us from them; and to note the velocity with which a storm is passing over the surface of the sun. See also "The Principles of Psychology," by Herbert Spencer, p. 617, etc. He says, "What is the constitution of this seemingly-simple matter, which thus tells of things near and remote that remain otherwise unknown? In the minutest visible fragment of it there are millions of units severally oscillating with unimaginable speed; and physicists show us that the amplitudes of their oscillations vary from moment to moment, according as the temperatures of surrounding objects vary. Nay, much more than this is now inferable. Each unit is not simple but compound—not a single thing but a system of things. Spectrum-analysis has made it manifest that every molecule of this so-called elementary substance is a cluster of minor molecules differing in their weights and rhythms. Such being the complexity of matters we lately thought simple, judge what is the complexity of matters we knew as compounds. In each molecule of an oxide or an acid, the chemist sees one of those systems united with one, two,

name for the unknown and hypothetical cause of states of our own consciousness? And what do we know of that "spirit" over whose threatened extinction by matter a great lamentation is arising like that which was heard at the death of Pan, except that it is also a name for an unknown and hypothetical cause, or condition, of states of consciousness? In other words, matter and spirit are but names for the imaginary substrata of groups of natural phenomena." *

three, or more systems of another kind that are similarly involved. Ascending to orders of compounds successively more heterogeneous, he finds himself obliged to recognise molecular complexities unrepresentable in thought; until, in reaching organic matter, he comes to molecules each of which (taking into account the composite nature of its so-called elements) contains literally more atoms than the visible heaven contains stars—atoms combined, system within system, in such ways that each atom, each system, each compound system, each doubly-compound system, has its motion in relation to the rest, and is capable of perturbing the rest and of being perturbed by them; * * * and molecules of each kind are specially affected by molecules of the same kind existing in the farthest regions of space. Units of sodium on which sunlight falls beat in unison with their kindred units more than 90 millions of miles off, by which the yellow rays of the sun are produced."

* Huxley "On the Physical Basis of Life."

# CHAPTER II.

## MAN.

It has been calculated that life began on this earth at least some 100 millions of years ago, while man is supposed to have lived here only 100,000 years. Professor Huxley, a great authority on this subject, says:—"The first traces of the primordial stock whence man has proceeded need no longer be sought by those who entertain any form of the doctrine of progressive development in the newest tertiaries; but they may be looked for in an epoch more distant from the age of the *Elephas primigenius* than that is from us."

Now how far it is *exactly* "*from* everlasting *to* everlasting" we do not know, and perhaps it would not be much less difficult to settle exactly the period here indicated—geological periods meaning only before or after certain occurrences; but we do know that millions of years before man appeared creatures lived on this earth which exhibited most of the mental phenomena displayed by him. They may not have been quite so pretty, judged by our standard of beauty,* but they were equally fond of their wives and families and friends,

* The inhabitants of the sea depart most from our type; the walrus, the norwal, the sea devil, or the cuttle fish are certainly not beautiful. But beauty is purely subjective, and even the female toad has her lovers and admirers, who are so ardent in their affections that Dr. Günter says he has often found the lady toad dead and smothered from their too close kisses and embraces. * * * In fact, there is no accounting for taste, either in sight or sound. Darwin says: "It is a curious fact that in the same class of animals, sounds so different as the drumming of the snipe's tail, the tapping of the woodpecker's beak, the harsh trumpet-like cry of certain water-fowl, the cooing of the turtle-dove, and the song of the nightingale, should all be pleasing to the females of the several species." ("Descent of Man," vol. 2, pp. 26, 67.)

and equally provided for their support and comfort; they feasted and fought, and crowed and blustered, and spread their tails to be admired, just as men and women do now; and really, when we come to compare an Australian Savage with Horses, Dogs, and Elephants, it is the animals, we are almost inclined to think, that ought to be ashamed to own the relationship. With reference to the vast superiority upon which man prides himself, and how infinitely ridiculous he makes himself in his self-importance, we should listen to the great philosopher, Michael Faraday:—"What a weak, credulous, incredulous, unbelieving, superstitious, bold, frightened—what a ridiculous world ours is, as far as concerns the mind of man! How full of inconsistencies, contradictions, and absurdities. I declare that taking the average of many minds that have recently come before me (and apart from that spirit which God has placed in each), and accepting for a moment that average as a standard, I should far prefer the obedience, affection, and instinct of a dog before it." And who does not recollect Byron's beautiful lines:—

When some proud son of man returns to earth,
Unknown to glory, but upheld by birth,
The Sculptor's art exhausts the pomp of woe,
And storied urns record who rests below;
When all is done, upon the tomb is seen,
Not what he was, but what he should have been:
But the poor dog, in life the firmest friend,
The first to welcome, the foremost to defend,
Whose honest heart is still his master's own,
Who labours, fights, lives, breathes for him alone,
Unhonour'd falls, unnoticed all his worth,
Denied in heaven the soul he held on earth:
While man, vain insect! hopes to be forgiven,
And claims himself a sole exclusive heaven.

Faraday is speaking of the Caucasian race, and not of the Mongolian, Malay, Negro, and Aboriginal American races, which at present, perhaps, comprehend five-sixths of mankind—

> But he who feels contempt
> For any living thing, hath faculties
> That he has never used.

The other sixth, the Caucasian, is much the highest type, and admits of a high degree of civilisation; and other races had probably existed on this earth many thousands of years before the yellow hair and white faces made their appearance. The most civilised claim, as Christians, the particular privilege of eternal damnation for the great majority; but we think it will be found that even they are not worth it, and that the whole plan of Creation will not be altered to suit the exigencies of their peculiar creed.

At present we have no data for fixing exactly the period of man's first appearance in the world. The Nile deposits, from which pottery has been brought up from a depth of 60 feet, and red brick from a depth of 72 feet, are calculated to represent 41,300 years, and it would be an awful bore, representing thousands of centuries, to get to the bottom of these deposits. No one has yet attempted to fix the precise date of the much-talked-of stone, or bronze periods, or of the lake dwellings; and fossil remains at present furnish very little that is definite as to time. Sir Charles Lyell is of opinion that the Engis Skull found in a cave in the valley of the Meuse is of the same age as the Mammoth and the Woolly Rhinoceros. The Neanderthal Skull is also said to belong to the same period, but no one has told us how long that is ago. In the time of flint axes and arrow heads the skulls found are small and round, with large ridges over the eyes. The race was small, and, it is believed, resembled the Lapps of the present day. In the bronze age the skulls were larger and longer. The skulls of the Egyptians, preserved in their tombs and temples, closely resemble the European, but are of smaller size.

The missing link between man and his fellow-creatures has not yet been found, but one-half the earth has not yet come under observation; and it is confidently asserted that, "between the highest primate and the lowest savage, there will be no break, no chasm, but a perfect and complete series." The Orang-Outang, found only in the Islands of the Indian Seas, probably presents as high a claim as any to be man's great progenitor. He is said to be as large as a man, and very like, with a physical force equal to eight men, and he is probably a less disgusting and degraded animal than the cannibal Papuan found in his immediate vicinity. It is highly probable, however, that Man's direct progenitor is buried under the sea in the Southern Hemisphere. Mr. Darwin says:—"The Simeadæ branched off into two great stems, the New World and the Old World monkeys; and from the latter, at a remote period, Man, the wonder and the glory of the universe, proceeded;" the early progenitors of man being no doubt once covered with hair, both sexes having beards, their ears pointed and capable of movement, and their bodies provided with a tail having the proper muscles.*

As regards the bony structure of the body, there is the closest anatomical resemblance between man and the anthropoid apes, every tooth, every bone being strictly homologous; but it is in the brain—the organ of the mind—that the greatest difference is found, and accordingly Professor Owen places man in a distinct sub-class of mammalia. It is not the peculiarities of mere physical form that most distinguish one animal from another, but the varied development of their nervous systems and the grades of mind dependent upon it. Thus, according to Leuret, the proportion of the brain to body is, in fishes, 1 to 5,668; in reptiles, 1 to 1,321; in birds, 1 to 212; and in mammals, 1 to 186.

The convolutions of the brains of apes, as of other animals,

* "Descent of Man," vol. 1, pp. 206, 213.

exhibit every stage of advance. The marmoset presents an almost smooth surface, while the orang has convolutions differing in appearance very little from man; the difference being more particularly in weight as compared to size of body. The lowest estimate for a healthy European brain is 32 ounces; that of a gorilla 20 ounces; while the latter is twice as heavy as any European of so small a brain. European brains have weighed 65 ounces, while the average is 42 ounces. The brain of Sir James Simpson, Edinburgh, according to the *British Medical Review*, was 54 ounces, and the average male brain, according to Quain and Sharpey, is 49½ ounces. The European brain is much heavier than the Hindoo, hence the supremacy of the former in weight of character. A man's brain is much heavier than a woman's, the average being as 42 ounces to 34 ounces; hence the dominance of man, which is not, as is often asserted, based on law or education. The capacity of the smallest human skull is, according to Mayner, 55·8 inches; that of the largest measured by Morton contained 114 inches,—the difference between the brain of the gorilla and the smallest brain of man being much less than that between the smallest and the largest healthy human brain. The whole nervous system of a man is twice as heavy as that of a gorilla; and "already the brain of the civilised man is larger by nearly 30 per cent. than the brain of the savage." *

"Idiocy results from an insufficient development of Brain; an adult male Brain being only nineteen inches in circumference necessitates mental imbecility, and all deviations from the normal shape of the Brain necessitate corresponding mental peculiarities. The circumference of the adult male head should not be under 21 inches at least, nor over 24 inches, the tape being passed over 'Philoprogenitiveness' and the Perceptive Organs (usually where the hat touches.)

* H. Spencer's "Principles of Biology," vol. 2, p. 502.

"The width of the male head varies from $5\frac{1}{2}$ in. to $6\frac{1}{2}$ in., measured with the calipers, from just above the ear.

"The height of the head should be exactly the same as the width, measured with the calipers from the opening of the ear to the vertical point in the coronal region. The length of the head should be one-third more than its width—that is, if the width be 6 in. the length should be 8 in.

"The tape measurement from the opening of the ear, at each side, over the coronal region, should be one-third less than the circumference, and the measurement from 'Individuality' over the coronal region to the occipital spine—the bottom of the skull at the back—should be the same as that from ear to ear.

"The female head is, as a general rule, from an inch to an inch and a half less in circumference than the male, and less also by an inch in width and heighth.. The posterior lobe of the female usually projects more than that of the male."*

"Physiologists have observed that each animal passes, in the course of its germinal history, through a series of changes resembling the *permanent forms* of the various orders of animals inferior to it in the scale. * * * Nor is man himself exempt from this law. His first form is that which is permanent in the animalcule. * * * The brain of man, which excedes that of all other animals in complexity of organization and fulness of development, is, at one early period, only (as quoted in 'The Vestiges from Lord's Popular Physiology,') a simple fold of nervous matter, with difficulty distinguishable into three parts, while a little tail-like prolongation towards the hinder parts, and which had been the first to appear, is the only representation of a spinal marrow. Now in this state it perfectly resembles the brain of an adult fish, thus assuming *in transitu* the form that in the fish is permanent. In a short time, however, the structure is become more complex, the parts more distinct, the spinal marrow

* "A Handbook of Phrenology," by Dr. C. Donovan.

better marked; it is now the brain of a reptile. The change continues; by a singular motion certain parts *(corpora quadrigemina)* which had hitherto appeared on the upper surface, now pass towards the lower; the former is their permanent situation in fishes and reptiles, the latter in birds and mammalia. This is another advance in the scale, but more remains yet to be done. The complication of the organ increases; cavities termed *ventricles* are formed, which do not exist in fishes, reptiles, or birds; curiously organised parts, such as the *corpora striata*, are added; it is now the brain of the mammalia. Its last and final stage alone seems wanting, that which will render it the brain of MAN. And this change in time takes place. It is to Dr. John Hunter and Tiedemann that we chiefly owe these curious observations." * Professor Owen begins still lower down in the scale, as he affirms that the human embryo is first *vermiform*, or worm-like. Thus, as the forms of life are mere stages of development, so are the different forms of brain and nervous system, and the grades of mind dependent upon them.

### THE FUNCTIONS OF THE BRAIN AND NERVOUS SYSTEM.

"Manifestations of intellect, feeling, and will, to every one but the subject of them, are known only as a transitory change in the body." † That is, Mind, objectively, like everything else, is known to us only by its mode of action. "Whenever much motion is evolved, a relatively-large nervous system exists; wherever the motion evolved, though not great in quantity, is heterogeneous in kind, a relatively-large nervous system exists; and wherever the evolved motion is both great in quantity and heterogeneous in kind, the largest nervous system exists." ‡ The nervous system

* "Vestiges of the Natural History of Creation."

† Huxley. ‡ Herbert Spencer.

presides over the motion, the brain over the direction or application of the motion. There are two kinds of nervous matter, white and grey; the white is connected with motion, the grey with its conscious direction. The body is covered with a net-work of nerves, thick in proportion to the sensibility required in the part. They have two functions, irritability and contractility—the one informing the mind of what is going on in the external world; and the other, embedded in the muscle, executing the orders of the mind in consequence. The one is called the afferent, the other the efferent nerve.

The nervous system is made up of bundles of threads enclosed in sheaths, which meet in local ganglia, the ganglia increasing in size until they meet in the great ganglion—the Brain. Those nerves that are acted upon by the outer world are in direct relation with those that react upon it, and these again are connected with those vital organs, whose function it is to accumulate the force which the others distribute.* "It is the transported shiver of bodies countless millions of miles distant," says Professor Tyndall, "which translates itself in human consciousness into the aspect of the stars" †; and the nervous system is highly-sensitive shivering stuff—"masses of unstable matter, undergoing decomposition with the greatest facility," through whose medium the shivering,

* "Considered as an agent for generating movements, the nervous system acts by liberation of successively larger amounts of molecular motion in the centres successively distributed. A very small change at the outer end of an efferent nerve sets up a relatively-large quantity of change in some adjacent unstable nerve-matter; whence the change, thus increased, is propagated to some internal ganglion; to be passed on by it immensely multiplied as before; until there is unlocked an amount of disturbance capable of causing muscular contractions throughout the whole body, * * * enabling it to make all parts of the body work together more efficiently in simultaneous and successive actions." (Herbert Spencer.)

† "Heat a Mode of Motion," p. 243.

vibratory, undulatory motions without us translate themselves "in human consciousness into the aspect" of the world; and we are told, on the high authority quoted above, that it would take 57,000 waves of the ether that produces violet colour to fill an inch; and as light travels at the rate of 192,000 miles a second, it would take 699 millions of millions of such waves to enter the eye in a single second to produce the impression of violet in the brain. "Star and nerve-tissue are parts of the same system—stellar and nervous forces are correlated, * * and matter and mind are bound together through measureless amplitudes of space and time." *

It is commonly supposed that thought travels faster than "the quick-winged arrows of light;" but this is a mistake. Thought travels at the rate of the nervous fluid, force, or agent in the nerves, and is dependent upon the molecular action set up in the brain. The rate of transmission, as measured by Professor Helmholtz, varies from 28 yards to 32 yards per second. In a rapidly-turning wheel we lose sight of the individual spokes, and the fire at the end of a stick when turned round rapidly appears to us as an unbroken circle, which shows that thought, or the transmission of sensation through the nerves, is less rapid than the motion of either the wheel or stick. So in hearing, a "tone" is made up of many separate sounds no longer distinguishable separately.†

* Dr. Youmans.

† A luminous impression lasts for about one-eighth of a second; whence it follows that if any two luminous impressions are separated by a less interval, they are not distinguishable from one another. Musical sounds at 16 per second are perceived separately. ("Psychology," pp. 108, 149, by H. Spencer.)

"Some remarkable experiments have been made in France on the relative rapidity with which messages pass through a telegraphic wire and impressions are conveyed through the nerves to the brain; and the result established is that the electric message travels with a

The Brain.—Mind is force—the highest development of force, "known only, to every one but the subject, as a transitory change in the body;" that is, like everything else, it is known objectively only by its "mode of action." Mental Philosophy is a pure system of dynamics, or measuring of forces. Dr. Gall, a physician of Vienna, was the first to begin the study of it on this method. Early in life he observed the different "modes of action" of his brothers and sisters and schoolfellows arising from peculiarity of talent or disposition. He soon began to connect external signs with these differences; to perceive that force of character was connected with a large head, and a narrow intellect with a narrow forehead. He also noticed, and was struck with the fact, that these narrow intellects had often good verbal memory, and stood high in the class from their great facility for learning by heart, and he observed that this peculiarity of talent was accompanied by prominent eyes. Afterwards, in the course of his professional studies, he discovered that this prominence of eye was owing to the comparatively large size of a convolution of the brain lying at the back of the eye, and which pushed it outwards and downwards. He then began to look for other mental characteristics connected with the brain, and whenever he found persons of very marked characters, either in feeling or intellect, if possible he took casts of their heads. He took casts of the heads of murderers, thieves, pugilists, of brave, or timid, or cunning people, of artists, sculptors, linguists, calculators, &c., &c., and he soon

swiftness many thousand times greater than the physical sensation, which is transmitted only at the rate of thirty metres in a second. Thus, supposing a whale thirty metres (about 103 feet) in length to be wounded in the tail, a second would elapse before the brain would be informed by the nerves of the event, and another second would expire before the will of the brain could be conveyed to the caudal muscles and action be taken to resent the injury, thus leaving an appreciable interval for the escape of the offender." (*Pall Mall Gazette*, June 16, 1869.)

found that these mental differences corresponded to differences in the shape of the head, which again corresponded to differences in the development of brain. By this method of proceeding, by tracing mental action to its source in nervous development, a very efficient and practical classification of our mental powers—of both intellect and feeling—has been arrived at, and their external signs can by competent, practised observers be now very correctly indicated,—force, that is force of mind, being here, as elsewhere, generally proportioned to the size of the structure or organ with which it is connected. Of course there are many other things besides size upon which the more or less perfect action of the brain depends, which we shall have to consider in their place.*

* The Physiologists of the present day generally leave off where Gall began; they have refused to follow his method, and have ignored his discoveries. Sir Henry Holland says: "Still there is one important fact here, which seems to be attested by as much evidence as the subject admits of. In every instance where there exists any corresponding lesion or disease on each side of the brain, there we are sure to find some express injury or impairment of the mental functions; and generally permanent, whatever be its peculiar nature. * * * Though absolutely unable to show, or even to conceive, how nervous matter should minister to these particular functions, or to others which it is better proved to maintain; yet if we found certain portions of it invariably connected with certain well-marked instincts, and never present where they are not, we might fairly affirm that we had gained some ground in advance." Speaking of Phrenology, he says: "The principle of division adopted, if principle it can be called, might have been carried greatly further."—("Mental Physiology," pp. 194, 232.) This is very true, but Gall was careful to confine himself only to what his observations warranted. He discovered about 26 organs. The last edition of "Combe's System of Phrenology" contains 37; and the Americans have subdivided or analyzed these, and added others, making now about 100. Finally, Sir Henry says: "In the present state of our knowledge of the brain, and of its relation to the mental functions, an impartial view of phrenology requires, not that the principle should be dismissed from view, but that great abatement should be made of its pretensions as a system. * * To say the least, it is chargeable with

Pursuing this method of observation, and classifying mind by what it does—by its different modes of action, the

what Lord Bacon has called 'an over-early and peremptory reduction into acts and methods,' and with the adoption of various conclusions, not warranted by any sufficient evidence. On the other hand, we must repeat the admission, founded as well on the complex structure of the brain as on the diversity of mental faculties and propensities, that the relation of parts to functions in this organ are more special and precise than any that have yet been unequivocally determined. And on a subject thus obscure, and where our actual knowledge is still limited to detached facts and presumptions, there is enough to justify the doctrine being kept before us, as one of the outlines to which future observations may legitimately apply;—not fettered, as they now are, by the trammels of a premature arrangement."—(Ibid, pp. 202, 210.) Sir Henry refers us to No. 17 of the *British and Foreign Medical Review*, where, he says, the argument for phrenology is very ably and fairly stated.

Dr. Henry Maudsley observes that "Psychology cannot be truly inductive unless it is studied objectively." He also says: "No one pretends that physiology can, for many years to come, furnish the complete data of a positive mental science: all that it can at present do is to overthrow the data of a false psychology. * * * The recognition of this increasing specialization and complexity in the function (mental) compels us to assume a corresponding development in the delicate organization of the nervous structure, although by reason of the imperfection of our means of investigation we are not yet able to trace a process of such delicacy in these most inmost recesses to which our senses have not gained entrance."—("The Physiology and Pathology of Mind," pp. 27, 165.) No doubt the Doctor will be delighted to hear that Dr. Gall, a physician of no less talent and eminence than himself, devoted his life to "tracing this delicate process," and that it is considered by *many* who have followed him that his life was not spent altogether in vain.

Professor Huxley tells us that "the functions of most of the parts of the brain which lie in front of the medulla oblongata are, at present, very ill understood; but it is certain that injury, or removal, of the cerebral hemispheres puts an end to intelligence and voluntary movement, and leaves the animal in the condition of a machine, working by the reflex action of the remainder of the cerebro-spinal axis.

"Thus there can be no doubt that the cerebral hemispheres are the

following tolerably complete list of its faculties and functions has been obtained:—

The SELF-PROTECTING FEELINGS, connected with convolutions of Brain lying at the base and sides of the skull.

The SELF-REGARDING FEELINGS, lying above the others at the back of the head.

The SOCIAL AFFECTIONS, at the back of the head, giving that elongated form peculiar to women.

The MORAL FEELINGS;

The ÆSTHETIC FEELINGS;

All these lying in the Coronal Region or upper part of the head.

FEELINGS which give Concentration, increased Power, or Permanence to the others.

The INTELLECTUAL FACULTIES, consisting of the faculties of direct and relative Perception and Reflection; these all lie in the front of the head, principally in the forehead.

seat of powers essential to the production of those phenomena which we term intelligence and will; but there is no satisfactory proof, at present, that the manifestation of any particular kind of mental faculty is especially allotted to, or connected with, the activity of any particular region of the cerebral hemispheres."—("Elementary Physiology," p. 304.) So that according to Professor Huxley we are only just where we were before Gall's time, nearly 100 years ago; all the laborious, careful, cautious induction of that observer have been thrown away, and the supposition that he had established the connection between certain mental faculties and particular regions of the cerebral hemispheres was all a fond delusion on his part, as it has continued to be of all his numerous followers since. I have, as will be apparent to all my readers, a great respect for Professor Huxley's opinion, and I do not know what weight he attaches to a "satisfactory proof;" but does he mean to affirm, for instance, that in the numerous cases of colour-blindness, where the organ of colour is deficient, or in cases of abnormal arithmetical calculating power, where the organ of number is in excess, there is no satisfactory proof of connection between the function and a particular part of the

Before we proceed to the consideration of the peculiar functions or modes of action of these faculties, it will be

cerebral hemisphere? This connection is easily verifiable by any one who chooses to look, and Phrenologists affirm that there is the same verifiable connection between mental faculty and all other parts of the brain; and this affirmation, to those who depend upon authority, is backed by a weight of testimony it is impossible to withstand, Professor Huxley's adverse opinion notwithstanding. George Combe, when a candidate for the Chair of Logic in the University of Edinburgh, in 1836, published a volume (Longmans and Co.) of distinguished names in testimony of the following propositions, viz.:—

"That Phrenology, viewed as the abstract science of mind, is superior to any system of mental philosophy that has preceded it;

"That it is a true exposition of the physiology of the brain;

Also, "To the application of Phrenology in discriminating the varieties of insanity;

"To its bearing on the classification and treatment of criminals;

"And to its application to the purposes of education."

As far as testimony is concerned, this book of the most distinguished names in this country and abroad may perhaps be considered a sufficient balance to the weight of Professor Huxley's unsupported opinion, as well as to the "contemptuous antagonism" which Mr. Herbert Spencer says phrenologists "have met with from both psychologists and physiologists." Mr. Spencer says: "That those who have carefully investigated the structure and functions of the nervous system, should have long ago turned their backs on phrenologists is not to be wondered at." I think it is greatly to be wondered at that the greatest names known to science in psychology and physiology should give their testimony in favour of phrenology in 1836, and that all who have carefully investigated *since* have turned their backs on it—unless, as the French physician said, "nous avons changé tout cela;" and as M. Camille Dareste has shown that transposition of viscera in the fowl can be brought about by the application of heat in a particular way to the egg, this may have been the case. I can only say, however, that I have never met with a single person who has "carefully investigated," according to Gall's method, who has turned his back on it. I have never even met with any of these contemptuous gentlemen who could map out a skull according to the phrenologists—that is, who knew what to look for in their investigations; and there can be little doubt that posterity will return with interest

as well to consider first upon what their general action depends.

the contempt with which they have treated Gall and the most important discovery the world has yet known.

Mr. Herbert Spencer does not so much object to Phrenology itself as to its present form. He says: "Saying nothing of many minor objections to the phrenological doctrine, we conclude that however defensible may be the hypothesis of a localization of faculties, when presented under an abstract form, it is quite indefensible under the form given to it by phrenologists."—("The Principles of Psychology," p. 576.) Every one seems disposed to give the poor phrenologists a kick: "hit him again, he has no friends," seems the very general cry. But, as Mr. Spencer says, "at best, Phrenology can only be an appendix to Psychology proper"; and "whoever calmly considers the question cannot long resist the conviction that different parts of the cerebrum must, *in some way or other*, subserve different kinds of mental action."—(Ibid, pp. 572, 573.) Could he not, in the Appendix to his "Psychology," now publishing, tell the world in what way that should be, and what *form* Phrenology ought to take? Phrenologists say that, by close and continued observation for about three-quarters of a century, they have traced the connection between certain modes of action of the mind—between certain feelings and intellectual faculties, and certain portions of the brain; that they have seen this connection thousands and thousands of times, the difference between a large and small organ being often from an inch to an inch and a half. Now, will Mr. Spencer say what *he has seen* that makes him think them all wrong, and that they have not seen what they say they have? Although casts are obliged to be definitely marked, yet phrenologists do not assert that very "specific limits" can yet be traceable in the brain; or that in the almost infinite variety of our thoughts and feelings their system of Psychology is yet complete. Phrenologists believe, with Mr. Spencer, in nervous centres: it has been admitted also, even by an opponent, that "their doctrine may now be said to be the only psychological one which counts any considerable mass of adherents."—(Lewes's "Biographical History of Philosophy," p. 628, library edition.) And valuable as Mr. Spencer's own system may be, it can scarcely be said to be as practical as the much-despised phrenological one, which in comparison is a "science of mind made easy." On this account alone his Phrenological Appendix to Psychology *proper* would be a great acquisition. Professor Huxley's and Mr. Spencer's present opinions on this subject,

LIFE.—The first requisite, of course, is Life, which, as far as we yet know, proceeds only from Life. At present it is only known to us as a force or vital spark transmitted from one body to another; a sort of innoculation, which gives to a body what is called Vitality. "A solution of smelling salts in water," says Huxley, "with an infinitesimal proportion of some other saline matters, contains all the elementary bodies which enter into the composition of protoplasm," (*i.e.*, living matter). * * * "An animal cannot make protoplasm, but must take it ready made from some other animal, or some plant, the animal's highest feat of constructive chemistry being to convert dead protoplasm into that living matter of life which is appropriate to itself." But if an animal cannot make protoplasm, neither can any chemical powers with which we are acquainted convert dead matter into living protoplasm. For, as Professor Huxley says, "Carbonic acid, water, and ammonia disappear, and in their place,

considering the amount of evidence collected, will no doubt in after times be classed among the hallucinations of men of genius.

Those who still prefer to look for living function in dead matter, and who refuse to use Gall's method, will probably look with considerable interest upon Dr. B. W. Richardson's late discovery. This gentleman—well known to science—says that in a dry human brain that he dissected he discovered membranes lining the convolutions that were really separable, and convolutions in which there were distinct centres.

Mr. Rutter has invented a magnetoscope, and by an improvement on this instrument by Dr. Leger, we are told by Dr. Ashburner that the magnetic force of each phrenological organ may be correctly indicated. "When the sums representing the organs of greatest activity were added together, the inference as to the character of the individual was easy, and seldom failed in being perfectly accurate."—("Philosophy of Animal Magnetism and Spiritualism," pp. 70, 71.) I have had no opportunity of verifying this discovery, but if true, as it reduces the question to one of simple addition, it may bring the investigation of the subject within the capacity of some of those physiologists and psychologists who now so contemptuously reject Phrenology.

*under the influence of pre-existing living protoplasm*, an equivalent weight of the matter of life makes its appearance." "Constructive chemistry could do nothing without this other *living influence*. Chemical powers are dominated by it, and only act freely when it is gone, *i.e.*, upon dead matter. Professor Tyndall, however, ageees with Huxley that "the tendency of modern science is to break down the wall of partition between organic and inorganic, and to reduce both to the operation of forces which are the same in kind, but whose combinations differ in complexity." Life is not what Herbert Spencer describes it to be, "The continuous adjustment of internal relations to external relations": these are only the *conditions* of its existence, and give no idea of what the transmitted or inherited "vital spark" is in itself. "Life," says Schelling, "is the tendency to individuation," *i.e.*, the forces of nature are confined within definite limits, and work towards a given object; and all vital actions have for their final purpose the elimination of nervous force, upon which mental force depends. The conditions of life, upon which the powers of mind depend, are the same in man as in all other animals.

FOOD.—It is from our food that the force or power that sustains vital action is derived. Sun-power divorces the oxygen from the carbon in the plant, and upon these elements coming together again in the bodily system an equivalent power is restored. The clash of atoms and arrested motion produce heat sufficient to work the whole system bodily and mental—that is, heat, known to us as a mode of motion, produces the peculiar mode of motion which characterises each organ, constituting what is called its function. And here let us pause to note an all-important distinction. When Heat is considered as a mode of motion, we are confounding the effect with the cause. Heat is not the mode of motion, but the cause, *i.e.*, the force which produces it. Much obscurity and error are occasioned by thus confounding cause and effect. For instance, it is falsely assumed that matter and motion *originate* force, whereas matter simply "conditions" force, or passes it on in a new form, of which motion is the sign; it never produces it; "for every manifestation of force must have come from a pre-existing equivalent force, and must give rise to a subsequent and equal amount of some other force. When, therefore, a force or effect appears, we are not at liberty to assume that it was self-originated, or came from nothing; when it disappears we are forbidden to conclude that it is annihilated; we must search and find whence it came and whither it is gone—that is, what produced it, and what effect it has itself produced." * Thus the body does not produce any of its forces, either Physical or Mental; the force is derived from the food, which is derived, either directly through plants, or indirectly through animals, from the sun.

No energy is *generated* by the bodily machinery; it can give out only what it has received. It passes through the body, working its various and complicated organs, and ultimately appears in its most compressed form as Consciousness or Mind—Mind being "the highest development of force, to whose existence all the lower natural forces are indispensably pre-requisite." † Throughout this work it must be distinctly understood that "force" represents an entity, not a condition. As motion cannot be separated and considered apart from the thing moving, neither can force be separated from that of which it is the force. The different names which we give to its different forms—of heat, light, electricity, chemical affinity, life, mind, &c., are dependent upon varying conditions of structure, and are known to us only in the different modes of motion of this structure, with one exception, and that is when it takes the form of mind or consciousness. In affirming, then, that "the Brain is the organ of the Mind,"

* "The Correlation and Conservation of Forces," by Dr. E. L. Youmans.

† "The Physiology and Pathology of the Mind," by Dr. H. Maudsley.

and that "Mind is the function of the Brain," precisely the same thing is meant. In neither case is it supposed that the peculiar force we call Mind is generated by the Brain. In the first case, its simple connection with the Brain is affirmed; in the latter, it assumes that the peculiar form in which the force manifests itself—its sensibility and mode of action—is dependent upon the Brain. This Force, or rather this unknown reality or entity of which it is the force, is the spiritual immaterial element throughout nature; it is the cause of all things, of all motion and emotion; for there is no reason to suppose that it is changed in its nature by passing through the brain. Its action is merely qualified by the matter—the atoms and molecules among which its energy is distributed. Its domain is not confined to merely physical phenomena; "the wondrous dynamic chain binds into living unity the realms of matter and mind." Here Materialism and Immaterialism meet

> As parts of one stupendous whole,
> Whose body nature is, and God the soul.*

Structure and Function.—What in inorganic bodies we call properties and qualities, in organic we call functions. Function is a power that a body displays, manifested in its mode of action. The peculiar mode of action depends upon the structure; the force or power does not; it is derived from some pre-existing equivalent force. From disregard-

* Dr. Bence Jones tells us, in his "Life and Letters of Faraday," that in that philosopher's "search for the unity of all forces he made all his great discoveries. Later in life a new image of matter came into his mind. He immaterialised matter into 'centres of force,' and he materialised the directions in which matter tends to move into physical lines of force. What he took from matter at its centres and gave to force, he partly gave back to matter in the lines of its motion." He might have made greater discoveries still if he had entered upon the study of Mind as Force, but to this phase of the subject he deliberately and, as I think, obtusely shuts his eyes.

ing this distinction there have been endless disputes as to whether structure always precedes function, or function structure, with about an equal amount of truth on either side. The brain and the muscles are increased in size by exercise, that is, by increased power supplied to the part. Here function is supposed to precede structure, for although the increased size or addition to the part or organ generally manifests the same function as the original, it does not always do so, hence the great *variety* in plant and animal. "Growth exceeding a certain rate, ends in the formation of a new aggregate, rather than an enlargement of the old." * Here we have a glimpse of the Law of Development; function always depending upon structure, but increasing force or power producing increased and, under peculiar conditions, varied structure, and consequently varied function. Every organ seems to have grown out of some other by a modification or adaptation rendered necessary by new conditions, or by some new purpose. Thus animals have been fitted to conditions, and conditions have helped to produce the animals. If we give time enough,—and there is plenty of time to give, and nature never seems to be in any hurry,—there is nothing inconceivable, or even improbable, in the supposition that all animal forms have arisen from successive modifications of one original form, however great the disparity may at present appear.† It is true it requires some

* "Principles of Biology," vol. 2, p. 27, by H. Spencer.

† To inquiring Negroes and Zulus, and other friends of Bishop Colenso, Noah's Ark, in which all living creatures were supposed to be assembled, has presented difficulties, and some naturalists even among ourselves have recently given expression to some doubts upon the subject; but we must bear in mind that it is quite possible that the true significance of *the fact* may not have been understood. Thus St. Augustine urges that the assembling of the animals in the ark must have been for the sake of prefiguring the gathering of all nations into the Church, and not in order to secure the replenishing of the world with life.—(See "Cosmos," vol. 1.)

study to see this, and it is flippantly said, "Surely man can never have passed along the oyster stage of being!" True, the divergence may have begun earlier, and man has probably passed along another line.

TEMPERAMENT.—Power or force is in no case generated by an organism, but is the exact equivalent only of that which is taken into it from without. The amount of force derived from the daily consumption of food in the full-grown body of a man, it is estimated, would raise 14,000,000 lbs. a foot high. This force, in the form of heat, is derived, as we have said, from the slow combustion of carbon within the system, in the same way as heat is derived from the coal that works a steam-engine.* This amount of force is required to work all the bodily machinery, and we must now add, what has been too long omitted, all the mental machinery too. The heart alone, contracting at the rate of 75 pulsations to the minute, during the 24 hours uses up about 500,000, or half-a-million, foot-pounds of this force, and the work of breathing has been estimated at 78,736 foot-pounds more. When heat is not generated in sufficient quantity, or when from exposure to cold the body is deprived of it too rapidly, we are all familiar with the consequences. Prolonged exposure depresses nervous action, and makes us sleepy; and hybernating animals, which as winter comes on,

* "On the constant supply of oxygen all fundamental power, and therewith the continuance of life, depends. The living body and the atmosphere around it constitute an inseparable whole. The once united elements still retain, in reality, their coherence—put asunder by force, and for temporary purposes, but pledged, as it were, to a deeper and inviolable union. In the re-uniting of the parted elements is effected the end and object of the whole process, the functions of animal life. Complex, wonderful, and beautiful as it is, surely the wonder and beauty of the organic world rise in this view to a yet greater height. For in the de-oxidation and re-oxidation of the hydrogen in a single drop of water we have before us, truly, so far as force is concerned, an epitome of the whole of life."—("Life in Nature." James Hinton.)

lose heat faster than they can make it, are deprived of motion and sensibility too, the heat being not sufficient to keep up that molecular action in the brain on which sensibility depends. If the heat is reduced still further, say to the temperature at which water solidifies, life and decomposition both cease. But this heat or force in its passage through the body is transformed in its character, and its "mode of motion" is changed by the various structures it passes through. As it was latent in the carbon and oxygen till they were brought together, so also in its action throughout the body it becomes alternately latent and active, and in this process much organic matter is used up, that which has parted with its energy losing its value and becoming waste, and requiring to be thrown out of the system. In its various transformations we have mechanical, chemical, electrical, odylic, vital, nervous, and mental force, the final purpose of all the others being, as has been said, the last; Mind, being the highest and most concentrated development of force, to the production of which all the other forces are indispensable. Each force, as it rises in concentration and importance, rises in the complexity of the structure required for its correlation or transformation; and mental power depends not only upon the size and perfection of its own instrument, the brain, but upon the degree of perfection in which all the other functions of the body are performed. The great desideratum is the proportional distribution of this food-force throughout the body, and the proper performance of each of its functions; and this depends upon the size and action of the organs or structures whose duty it is to perform those functions. The predominance in the system of particular organs, or set of organs, has given rise to what have been called the Temperaments. Thus we have in some constitutions the brain and nervous system predominating; in others the blood-vessels; in some the muscular system, or the stomach, glands, and assimilating organs.

"The different temperaments are indicated by external

signs, which are open to observation. The first, or *lymphatic*, is distinguishable by a round form of the body, softness of the muscular system, repletion of the cellular tissue, fair hair, and a pale skin. It is accompanied by languid vital actions, with weakness and slowness of the circulation. The brain, as part of the system, is also slow, languid and feeble in its action, and the mental manifestations are proportionally weak.

"The second or *sanguine* temperament is indicated by well-defined forms, moderate plumpness of person, tolerable firmness of flesh, light hair inclining to chestnut, blue eyes, and fair complexion; with ruddiness of countenance. It is marked by great activity of the blood-vessels, fondness for exercise, and an animated countenance. The brain partakes of the general state, and is vigorous and active.

"The fibrous (generally, but inappropriately termed the *bilious*,) temperament, is recognised by black hair, dark skin, moderate fulness and much firmness of flesh, with harshly-expressed outline of the person. The functions partake of great energy of action, which extends to the brain; and the countenance, in consequence, shows strong, marked, and decided features.

"The *nervous* temperament is recognised by fine thin hair, thin skin, small thin muscles, quickness in muscular motions, paleness of countenance, and often delicate health. The whole nervous system, including the brain, is predominately active and energetic, and the mental manifestations vivacious and powerful." *

Of course these temperaments are combined in various proportions in different constitutions, but they all have their external signs, and the degrees in which they are possessed are readily discernible. As they act upon the mind, so does the mind re-act upon them, and the relative proportion in which they exist may be much modified by exercise and

* George Combe's "System of Phrenology," 5th Edition.

treatment. Still they furnish but a rough approximation to the bodily influences affecting the action of the brain and mind.

The first requisite for mental action is sufficient force to produce molecular motion in the brain; for if this is deficient in quantity, as shown by the want of heat, or from the force being engaged elsewhere, as in sleep, total or partial insensibility is the consequence. If molecular action is interfered with by pressure, we have also the same result.

The temperaments act principally through the quantity and quality of the blood which they supply to the brain, as they regulate its rate of circulation and its elementary components—more or less rich in the constituents of nerve-substance. This fact requires much more study than has yet been given to it. We know how temporarily to increase the circulation, but the special food the brain requires to produce the greatest evolution of nerve-force has yet received very little attention. If the feeders to the brain are imperfect, and it is scantily supplied with nervous force, or if there is an unusually large brain with only the ordinary support, that is, the brain is larger than its feeders, or if from a very active temperament there is more than the usual expenditure of mental force, great depression of spirits is the consequence, in one case constant, in the other intermittent.

Dr. Robert Bird, of the Bengal Army, in a volume of Physiological Essays, just published (Trübner and Co.,) has helped to throw considerable light on this subject. He shows how the slightest modification of tissue makes a difference in the quality of brain, and makes one man mentally energetic and distinguished, and another mentally lazy and obscure. The agencies by which the tissues of the body are charged through life, are use, social condition, climate, mixed parental influence, and hereditary transmission, and as through physical excellence it has been ordained that we should attain unto moral excellence, men must necessarily progress towards perfection, because the physical universe is changing and daily becoming fitter for the growth of a higher race of beings. As regards the requisites for healthy tissues, he says: "The same tissue is differently moved by different external agents. Nay, each tissue appears to be moved by certain agents, and by other agents not to be moved at all. From this I conclude that the different tissues and the different modifications of the same tissue have affinities to certain elements of our surroundings which they have not to the other elements. When it happens, then, that those elements which only have the power to move a certain tissue, are withdrawn, that tissue, for lack of exercise, begins to waste. By reference to this principle, I explain the phenomenon of a man becoming idiotic while undergoing a course of solitary confinement. In his cell he may have enough of food and fresh air, and he may be kept clean and comfortably warm. But cleanliness, sufficient food, fresh air, comfortable warmth, and even light, do not make up the sum of conditions necessary for the maintainance of perfect health. He has within him tissues and modifications of tissue which require for their welfare the waving of woods and corn fields, the scent of flowers, the gleam and ripple of running water, the battling of wind and rain over plains and hills, the struggle with our fellow-creatures for life and position, the ferment of crowded cities, all the sweet variety of domestic life, and much besides: denied them, they fret and waste, and ceasing to influence the economy to which they belong, they abandon the poor prisoner to gloom, indifference, idiocy."* No doubt this is sufficient to explain much of the benefit that is said to be derived from "change of air," but idiocy is caused by atrophy of the intellectual faculties only; the derangement of the emotions would cause, not idiocy, but insanity. Dr. Bird does not at present appear to have apprehended that the Intellect and the Emotions are connected with different lobes of the brain, but speaks of the mental capacity

* "Physiological Essays," p. 129.

generally as being in proportion to the size and quality of brain (see p. 76). Dr. Bird believes, he says, "that the greater or lesser tendency which our diseased tissues show to take on diseased conditions is the result of all the combinations of external influences to which we in our own persons, and in the persons of our ancestors, have been subjected previously to the time when we become diseased."* And on this account I suppose it is that the same medicines act so differently on different constitutions. Thus Dr. Bird tells us that "it may be stated broadly that everything external to us affects us beneficially or prejudicially according to the condition of our tissues at the time when it affects us, and to the measure in which it is applied. Too little food is in its results scarcely more disastrous than too much food, and the same may be said of all stimulant substances. Then all the substances named in the "Pharmacopœia" are medicinal or poisonous, according to the state of our tissues when they are used, and to the measure in which they are used. * * * Indeed, it is the rule that no two bodies are acted upon in the same degree by the same dose of medicine." † ‡

By carefully-prepared tables Dr. Bird shows that the specific gravity of Europeans is greater than that of Indians—natives of Bengal. There is a higher degree of heat indicated by the thermometer in the mouth. He mentions other differences, and says men will have to be classed not always according to their nation, but according to the nature of the

* "Physiological Essays," p. 221. † Ibid, pp. 109, 187.

‡ Hobbes drank cold water when he was desirous of making a great intellectual effort. Newton smoked. Bonaparte took snuff, Pope strong coffee, Byron gin and water. Wedderburne, the first Lord Ashburton, always placed a blister on his chest when he had to make a great speech. The great Lord Erskine took large doses of opium. On the trial of Queen Caroline, Erskine, anxious to make a great speech, took an overdose of his favourite drug. The effect was striking; he dropped into the arms of Lord Stanhope, who sat next him.—*Medical Times and Gazette.*

materials of which their bodies are composed.* In the Bengalee the analysis of blood shows more albumen and fewer red corpuscles than in the European; so that a man is strong, I suppose, according to the quantity of iron in him. These red corpuscles appear to be very important, since they increase as we rise in the animal scale, until they attain their maximum development in carnivorous animals and birds. They increase with an improvement of the breed in animals, and they decrease with old age, and in those shut out from light and fresh air, and they are less in women than in men.† The blood from young people keeps sweet longer than

* Dr. Lankester, in the pages of *Nature*, has lately drawn attention to the value which should be attributed to weight in comparison with height when testing the soundness of the human frame. It is a well-known fact that the tallest soldiers are less able to bear fatigue and exposure than those whose height is rather below the average, and that the instances are very rare in which longevity and great stature have been combined. From investigations made by Dr. Hutchinson and Mr. Brent, it appears that a person five feet high ought to weigh 8st. 5lb., and that a healthy man increases for every additional inch of height five pounds in weight; whenever any considerable divergence from these rules is observed disease may be suspected.

† It is not generally known that John Wesley set up for the cure of bodies as well as of souls, and that he approved himself quite as much in the former capacity as the latter, although knowing, as we now do, the value of the blood, not, perhaps, with sufficient reason. Many, however, now think his views of both equally well grounded. He was not a little proud of his "Primitive Physick, or an Easy and Natural Method of Curing Most Diseases," printed by William Pine, in Narrow Wine Street, Bristol, and sold at the New Room in the Horse Fair, and in London, 1762. Wesley announced that "every man of common sense (unless in some rare cases) may prescribe either to himself or his neighbour, and may be very secure from doing harm where he can do no good.". Among the remedies which he approves as "tried"—a word which he thus made proverbial in the Methodist connection—is bleeding for consumption. The patient is to lose six ounces of blood every day for a fortnight if he live so long, and then every other day, then every third day, then every fifth day for the same time. The gout is to be cured by the application of a raw lean beef-steak; for twisting of the bowels, one, two, or three pounds of quicksilver in water.

that taken from the old, which shows a gradual advance towards decay and dissolution even in the living. Dr. B. Richardson measures this decreasing strength: "in the form of heat," he says, "we have a measurement of the capacity of the body to sustain force, which is only another phrase for expressing the resistance of the body to death. For example, if we assume that a healthy man of 29 years respires 450 cubic feet of air per day, and by combustion of his carbon evolves as much heat as would raise 50 lbs. of water at 32 deg. Fahrenheit to 212 deg. Fahrenheit under all ordinary temperatures, which is about the fact; and if we assume that another man at 39 shall not be able at any temperature to respire so much air, and shall not be able to evolve as much caloric as would raise 44 lbs. of water from 32 deg. to 212 deg., we see a general reason why the latter man should feel an effect from a sudden change in the temperature which the younger man would not feel," &c., and so on in proportion, till at 81 a man raises only eleven pounds.*

The fact has not yet been sufficiently attended to that there is no spontaneous generation of force anywhere, and certainly therefore not in the human body. The contractor was quite right when he said, "If you don't put a lot of stuff into a navvy you can't get nothing out of him." The amount of force we receive from food is a definite quantity; it is all there is to do the work of the body, and if it is used up otherwise, it cannot be used in thinking and feeling. Some food may be very nutritious, but if it requires a great power of digestion, we are little the better for it on that account, as it requires almost as great an expenditure of force as it gives. Thinking and digestion do not go on well together, because the force required for active thinking is doing duty in the stomach, and if we insist upon using it in the brain, digestion

* "Waves of Heat and Waves of Death," *Popular Science Review*, January, 1865.

is impeded. So of the expenditure of force in muscular exercise: much out-door work is incompatible with vigorous thinking, and if we have a large body of our own to carry about we require a proportionally larger brain. The same may be said of every function of the body. If a large amount of force is used in its exercise, there is less to convert into sensibility of any kind. We see, then, how extremely important it is that no force should be allowed to run to waste, that all the machinery should be kept in proper working order, all the joints well oiled. The animal body is by far the most perfect machine that has yet been made for the economical use of force. "Of the total heat given out by the combustion of the food, a man can make a fifth available in the form of actual work, while it has never been possible to construct a steam-engine that could utilize more than a ninth of the energy of the fuel burnt under the boiler. But in addition to this external work the body has constantly to perform a vast amount of work in order to sustain the life. There is the blood to be kept circulating and urged through the lungs and capillaries; the chest and diaphragm have to be raised for the purpose of breathing; digestion has to be carried on, and the body kept erect—all these consuming energy." * The brain, or thinking power, has not yet been included by physiologists among the force consumers; and yet it is the most important and the most expensive of them all. There is nothing so exhausting as violent pain, or strong feeling or emotion; and, in fact, every thought or feeling requires an expenditure of what we call physical force. As we have said, then, the great desideratum is a large generation of force and the proportional distribution of it throughout the body, each organ performing its proper function; for if one predominates it is at the expense of the others. Thus we may have mental power at the expense of the vital, and muscular or locomotive at the

* "Animal Force and Animal Food," by J. Broughton.

expense of the mental. Genius, dependent upon great activity of brain, often absorbs too large a proportion of the vital power, at the expense of health. Genius, to be safe, should be connected with a large brain, and a proportional development of body. The causes of the great power, which some constitutions possess, of bearing fatigue, of longevity, of genius, of the *vis medicatrix*, &c., are as yet quite mysteries to us, but they will probably be found to depend upon the mode and proportion in which the food-force distributes itself; and this depends upon the speciality of structure.*

* Mr. George Combe, in his "System of Phrenology," vol. 1, p. 294, mentions that a physician in Philadelphia, Dr. George McLellan, saw reason to believe that tenacity of life bears some relation to the development of the organ of "Vitativeness." Dr. C. Donovan ("Handbook of Phrenology," p. 120) says that his experience is in favour of Dr. McLellan's notion that the condition of the skull as to width in the basilar region (about the ears) and corresponding vitality is correct. He also says:—"On the size of the Brain root, the '*medulla oblongata*,' and on the size of the spinal column in general—for this will ever quadrate with the medulla oblongata—the degree of vital power and tenacity of life, apart from the Love of Life, depends. The foramen, or round hole through which the spinal column enters the skull, will be large or small in accordance with the size of the Brain root. That this foramen varies in size in different skulls is obvious. To come to the external indications of the size of the Brain root, we have to refer to the root of the ear. This in some persons is in a hollow or pit, as if a tablespoonful of matter had been taken out to let the root of the ear be implanted, whilst in others the ear stands out on a level with, or even beyond, the skull. In order to the better understanding of this important point, let it be supposed that a modeller in forming the human head had omitted to place the ear in its proper position. Let it be supposed that in supplying this omission he were to scoop out about a tablespoonful of clay in the side of the head, and place the ear in the hollow thus made. This would give the ear a narrower space from side to side, and would give it a *pitted appearance*. Take the reverse case, and let the modeller, instead of making a hollow to plant the ear in, place *it on the even surface*, or even on a slight eminence, so as to cause it to

RACE.—It seems most probable that the five races, as distinguished by their colours of White, Black, Brown, Red, and Yellow—the Caucasian, Negro, Malay, American Indian, and Mongolian,* were not derived from one stock, but had each a separate origin; the causes that could produce a single man being quite equal to produce all the varieties. Life has been described to be "the continuous adjustment of internal relations to external relations," and this, if not life, is at least the condition of its existence; and if we consider that this "adjustment" has been going on for perhaps a million years before the forces that constitute the bodily and mental constitution of a man could be brought to act in their present

protrude a little. In the first case the pitted ear would indicate a low state of vitality, whilst the second would represent an active and vigorous state of vital power. * * * Granting the soundness of the pitted ear theory, it would appear that in the study of mental character the size of the 'medulla oblongata,' *in which the cerebral nerves are rooted*, and the degree of prominence of the ear caused by such size, is a question of great importance as regards both physical and mental vigour. In a medical point of view this question is of very great consequence. Persons with low vitality when in good health are obliged to eat frequently, because they cannot effectually *extract due nutriment from their food*. When the appetite fails they have recourse to fluids to sustain them. If, when ill, they are kept on low diet, they are apt to slip through the doctor's fingers. All the long-lived have the ears well out on the head."

* According to Prof. Huxley, the five distinct types of Mankind are the *Australioid*, found in Australia, in the Dekhan, and formerly in the valley of the Nile; the *Negroid*, including the Negroes and Bushmen of Africa and the Negritos of New Guinea, Tasmania, &c.; the *Xanthochroic*, distributed through Iceland, Eastern Britain, Scandinavia, North and Central Germany, and extending through Eastern Europe into Asia as far as North-western India, and found also in North Africa; the *Melanochroic*, located in an area situated between the Xanthochroic and Australioid peoples; and the *Mongoloid*, a large and somewhat ill-defined group occupying Central and Northern Asia, the two Americas, and Polynesia.—"*On the Chief Modifications of Mankind, and their Geological Distribution.*" *A Paper read before the Ethnological Society, June 7th*, 1870.

persistence and harmony, we may then get a slight idea upon what the stability of type and race depends. It takes a long time to make any two forces act unconsciously or automatically together, and the forces that constitute the harmony of the human system are infinite. Forces thus made to act together to effect a definite purpose are not easily disunited. If, then, the different races of mankind had a separate origin, and the forces of which they are composed were brought together and associated by a different environment, "the structure of an organism being the product of the almost infinite series of actions and reactions to which all ancestral organisms have been exposed," * then such races could not be expected to mix with advantage. Automatic actions would clash, and there would always be the tendency, as in fact exists, to return to the original types. Thus races which had probably the same ancestral types, such as those that now inhabit Europe, are crossed with advantage; but we obtain only mongrels, inferior perhaps to either original, from the crossing of distinct races. Darwin says: "The first meeting of distinct and separate people generates disease." He also says: "If we look to the races of men as distributed over the world, we must infer that their characteristic differences cannot be accounted for by the direct action of different conditions of life, even after exposure to them for an enormous period of time. The Esquimaux live exclusively on animal food; they are clothed in thick fur, and are exposed to intense cold and to prolonged darkness; yet they do not differ in any extreme degree from the inhabitants of Southern China, who live on vegetable food, and are exposed almost naked to a hot, glaring climate. The unclothed Fuegians live on the marine productions of their inhospitable shores; the Botocudos of Brazil wander about the hot forests of the interior, and live chiefly on vegetable productions; yet these tribes resemble each other so closely that the

* Herbert Spencer.

Fuegians on board the "Beagle" were mistaken by some Brazilians for Botocudos. The Botocudos, again, as well as the other inhabitants of tropical America, are wholly different from the Negroes who inhabit the opposite shores of the Atlantic, are exposed to a nearly similar climate, and follow nearly the same habits of life." *

Mr. Darwin, nevertheless, rather leans to the opinion that we had but one ancestor. Mr. Wallace, on the other hand, holds that man passed through more than one channel of derivation, or transitional form, from the class of the inferior mammals; such primary differences as those of Negro, Caucasian, or Australasian, denoting the special strain or breed of quadrumana from which each is supposed to have risen to the dignity of man. The colour of the skin depends on race more than on climate. It "depends on the pigment excreted from the blood, and interposed between the cutis and cuticle. This seems to be excreted from the surface of the cutis, and, like the cuticle, is extravascular. It is as thick as the cuticle in the negro, and, by nice manipulation, can be detached as a separate membrane. According to dissections made by Sœmmerring and Hunter, the texture of this intermediate lamella between the cutis and cuticle exists in the fairest of Europeans; but the pigment is not deposited, and hence the colour depends on the transparency of the skin revealing the blood in the capillaries beneath. From white to black, every conceivable shade is produced by the amount of colouring matter deposited in this lamella or rete-mucosum." †

As Malthus has shown, the world may be filled from the good stock with which now it is everywhere inoculated, and the inferior races, although they at present represent five-sixths of mankind, on the principle of Natural Selection, will probably be "civilised" off the face of the earth.

That he is a progressive animal is the characteristic dis-

* Vol. 1, pp. 239, 246.

† "Physical Man," p. 129. By Hudson Tuttle.

tinction of man, but several of the inferior races of man seem quite incapable of progress; the Black, Brown, and Red do not appear to have advanced a step so far as we have any record. From ages of locomotive hunting habits, the vital and muscular forces so predominate in the American Indian, that any lengthened rest would be to him painful, and any settled mode of life impossible; whereas ages of patient industry have as much fitted the Chinese to do the work of the world as constant locomotion has disqualified the Indian. In the East civilization seems to have made considerable progress, and then to have become stereotyped. He who discovers the cause of this stereotype, and by removing it sets civilization going again, may perhaps save one-third of mankind from extermination. Cæsar's description of the people of his time, contrasted with the little change their descendants have undergone during the last thousand years, remarkably illustrates the permanence of type, and proves that in the institutions we would give to a people the first thing to consider is Race. England has failed to govern Ireland, and the Emperor of the French failed in Mexico from not sufficiently considering the race he had to deal with.

HEREDITARY DESCENT.—We have seen how much depends on Race, but quite as much depends on hereditary descent. "He comes of an honest family" is proverbial among the poor, and it is astonishing how little deviation there is from this rule. The same may be said of roguery—"What is bred in the bone cannot be got out of the flesh." The microscopic germ contains the red hair and other physical and mental characteristics of the father, however modified afterwards by the character of the mother, and her treatment during the time of gestation.* Why in some families the characteristics

* It is a fact, settled by innumerable observations and experiments, that the bees can so modify a worker in the larva state that, when it

of one parent entirely obliterate those of the other, or on what principle they are mixed, is not yet known; but in judging

merges from the pupa, it is found to be a queen or true female. For this purpose they enlarge its cell, make a pyramidal hollow to allow of its assuming a vertical instead of a horizontal position, keep it warmer than other larva are kept, and feed it with a peculiar kind of food. From these circumstances, leading to a shortening of the embryonic condition, results a creature different in form, and also in disposition, from what would have otherwise been produced. Some of the organs possessed by the worker are here altogether wanting. We have a creature "destined to enjoy love, to burn with jealousy and anger, to be incited to vengeance, and to pass her time without labour," instead of one "zealous for the good of the community, a defender of the public rights, enjoying an immunity from the stimulus of sexual appetite and the pains of parturition; laborious, industrious, patient, ingenious, skilful; incessantly engaged in the nurture of the young, in collecting honey and pollen, in elaborating wax, in constructing cells and the like! paying the most respectful and assiduous attention to objects which, had its ovaries been developed, it would have hated and pursued with the most vindictive fury till it had destroyed them!"—(Kirby and Spence). The bees were inhabitants of this earth long before man, and have had, therefore, a much longer experience, and when we have been as long in the world perhaps we shall take a lesson from them. When civilisation no longer requires the pressure of population on the means of subsistence, then we may turn our attention to making queen bees and neutrals. At the present time even there are "advanced" American ladies who cannot be said to be "true females."

Louis Figuier, in "The Insect World," tells us that "the working-bees, unfruitful females, with a self-denial very rare in nature, seem to have no other vocation than to sacrifice themselves to the welfare of the larvæ." Might not the strong-minded ladies who are now standing up for the "Rights of Women," and seeking work, find their occupation here in the development of the human larvæ, and console themselves with the reflection that such a condition is higher than the mother's, and requires a more advanced growth both in body and mind, taking four days longer to develope from the larvæ? "Workers" at home, and the education of the rising generation, that is their department, when properly understood. It is stated that in Philadelphia, out of 1,194 teachers, there are 1,110 women and 84 men teachers; while in New York, out of 26.000 teachers, there are 21,000

of the transmission of hereditary tendencies, we must have both genealogies before us. This transmission is readily accounted for—exercise augments function, this increases structure, and this increased structure is transmitted. "Hereditary tendencies can be ascribed to nothing but inherited development of structure, caused by augmentation of function." * Strong impressions, which alter structure, and which leave their mark upon the easily impressible substance of the brain, are transmitted, and thus most peculiar idiocyncracies run in families, showing themselves often at the period of life in which they first appeared in the parent. "Some fault, some contravention, some ignorance of the eternal laws—who knows how, or how long ago committed?—has entailed this Nemesis." †

Sir J. Herschel says: "The disregard and ignorance of the laws of human organisation manifested in the transmission of diseases to posterity deserves the severest censure, and often receives the severest punishment," and, may we not add, that the transmission of mental disease is equally culpable.

women and 5,000 men instructors."—(*Athenæum*, June 2, 1870.) So greatly does instinct go ahead of reason in America, where all the talk is of the "Rights of Women" to become men.

Again, might not our fast young ladies, as well as the "sluggard," go to the ant and take a lesson, as well as from the bee? In the tribe of ants "some brilliant black flies, with their delicate wings in continued motion, make their appearance among the workers. These are the male and female population of the colony, which must some day face abroad, and go to establish new families at a distance." These winged ants swarm out into the bright sunshine, live for a time a rather fast life, each seeking for one to love. Having made their choice, no doubt a happy one, they settle down to the duties of maternity, and mark! they then tear off their wings "by waving them, so as to weaken the muscles which unite them to the corselet, and pinching them off with their legs," that they may no longer be tempted to roam abroad and neglect their household duties.

* Herbert Spencer. † T. A. Trollope.

Climate.—Whether climate is more important than race, or whether race itself grows out of climate, are still disputed points with ethnologists; but surely the immense contrast between the Arab and the Negro, both in body and mind, is not sufficiently accounted for by the difference in the climates which they respectively inhabit. Still climate, it must be allowed, is all-important. We can make little progress till we give our mental and moral philosophy a physiological basis, and, physiologically, man grows out of the earth as much as the plants do. It is in vain that he tries to set himself up as something distinct and apart from nature, whereas he is a part only of the great whole, his present always growing out of the past; the rudiments of all his highest acquisitions being found in the strata below. The sun's spots even, and variable influence, affect the climate and atmosphere, and through them the vegetable kingdom, which again influences both animals and men, changing tissue and its mental manifestations. The sun itself is under the influence of other bodies in space, so that the remotest star acts directly on our organisation, and we may yet have to admit the astrologist within the boundaries of science, and be glad to know that we are born under a favourable star.

Differences in climate, Dr. Bird tells us, may be referred to differences in heat, moisture, chemical action, electricity, magnetism, and light; and differences in these may be referred to differences in distance from the sun, moon, and stars, from the equator, in the relative distribution of land and water, in the irregularity of the earth's surface, in the composition of the earth's crust, and probably in differences in the relations which exist between different portions of the earth's crust and the interior of the earth, where the elements of volcanic action are supposed to reside. Men take their characters from contact with the physical influences among which it has been their lot to be cast. Nations and tribes take their places in the human family according to their physical good fortune or their physical bad fortune; the inhabitants of swamps and jungles are necessarily of lower organisation than are the inhabitants of breezy and well-cultivated uplands. Humboldt, in his "Cosmos," and others have now made these things pretty well understood; but that man is a great magnet, and sympathises with the varied magnetic and electric condition of the earth is less known. Changes are constantly taking place in the intensity of terrestrial magnetism. Thus it is increasing in America, and decreasing in Europe and Asia; and when from the general we descend into the particular, we find that at Toronto, in Canada, the total force in the dip inclination in the magnetic needle increases annually 1·0, and that in London there is an annual decrease of the same of 2·7. This shows, says Dr. Bird, that a benefit or a loss is travelling westwards. Can it be true, he says, as the Yankees continually tell us, that Europe is growing old; that, having borne the burden of a high civilisation for thousands of years, she is transferring it to the back of the unexhausted energies of America?* It certainly admits of a question as to whether Europeans are improving in America, as to whether there is a "benefit or a loss?" Anthropologists say: "We have exhaustion and degeneracy, but no real acclimatation."† In the American the nervous system greatly predominates. This Nathaniel Hawthorne ascribes to the dry atmosphere and vicious habits of life; but both the vicious habits and nervousness may quite as likely be owing to the difference in the magnetic condition. "John Bull, on the other hand," says Hawthorne, "has grown bulbous, long-bodied, short-legged, heavy-witted, material, and, in a word, too intensely English." Let us hope that both extremes may be modified by food, training, and proper educational treatment.

In considering this subject we must also not lose sight of

* "Physiological Essays," pp. 144, 119, 125.

† "Anthropological Journal," July, 1868.

chemical action and the consequent evolution of electricity, not only dependent on winds and waters on the surface, but upon the chemical action going on in the crust, and in the bowels of the earth. Faraday says that the chemical action of a grain of water on four grains of zinc can evolve a quantity of electricity equal to that of a powerful flash of lightning. As chemical action is daily going on to a vast extent in trees, and in all kinds of vegetation and in animals, a large proportion of the atmospheric electricity must come from this source. The production of electricity varies greatly in different localities from these and other causes, giving new properties to matter, and consequently a different action on, and relations to, our tissues, and we begin to understand how it is that many changes take place which the thermometer cannot measure; and how the heat of one wind acts differently from the heat of another wind; and how people should exclaim, How depressing this wind, and how exhilarating that! *

The action going on in the earth's crust generates all kinds of peculiar influences, which greatly affect some people, and all differently. Some are affected by metallic currents, some by running water, others in passing over graves, and in various other ways. Finding metals and streams of water with the hazel wands, by people with abnormally constituted nervous systems, has passed into a profession, and hundreds of volumes have been written on these phenomena of Rhabdomancy, as they are called. The emanations of these mundane agents are not confined to veins of metal, or subterranean streams, but the whole mineral kingdom has a profound and mysterious relation with the organism of man, and different constitutions are thus differently affected by different localities. At Delphi we have prophecies, in the Black Forests ghosts and demons. The earth may be made into a battery; man is a magnet or a telegraphic wire,

* "Physiological Essays," pp. 200, 203, 206.

oftener, however, a post, and the electric currents that circulate freely through the different strata, often mixed with decomposed organic matter, affect him accordingly.*

This question of climate, and especially of mundane influences on the tissues and brain-centres, and even on health and on disease may be said to be quite in its infancy. I give the following, however, as illustrations of the attention which is beginning to be devoted to the subject:—

At a meeting of the Geological Society a paper by Mr. Whitaker was read "On the Connection of the Geological Structure and Physical Features of the South-east of England with the Consumption Death-rate," in which a fact long insisted upon by sanitary reformers was established from the geological point of view. In fifty-eight registration districts of Kent and Sussex consumption most prevails where the soil is wet. This result, however, does not depend entirely on the stiffness or porousness of the soil, for it is modified by slope and elevation. It has long been remarked in Devonshire that consumption prevails in the valleys and not on the hills. Is it quite certain that the hollows are warmer than the heights? The country-folk round about Haslemere all declare that the cottages on the hills are warmer than those in the valleys; and the same testimony may be heard in other parts of England.

Dr. Moffat also attempts to establish the close connection between certain soils and certain maladies. His experience has chiefly lain in Cheshire, and he says he has there found that consumption and anæmia follow in a very remarkable manner the course of the carboniferous system, and prevail to a far less extent among the dwellers in the new red sandstone district. So far this agrees with the above observations of Mr. Whitaker and with what Dr. Buchanan and Mr. Simon observed in the counties of Surrey, Kent, and Sussex,

* See "Philosophy of Mysterious Agents," chap. 5, by E. C. Rogers. (Jewett and Co., Boston.)

where phthisis was most fatal upon the damp soils; and it is also supported by the fact that in large towns the disease has been the first to decline in virulence after the ground has been rendered dry by the introduction of main sewers. The novel parts of Dr. Moffat's theory is, however, that analyzing the wheat grown upon the two geological systems, he found that in that produced upon the sandstone there was a large proportion of phosphoric acid and of oxide of iron, while in that of the carboniferous system these constituents were almost wholly wanting. He concluded therefore that the prevalence of consumption and anæmia were mainly attributable to the deficiency of these elements of blood in the ordinary diet of the population.

"HARD WATER.—The testimony of public waste is decidedly in favour of hard water. It is usually rendered hard by the suspension in it of calcareous substances, which are rendered soluble by the presence of carbonic acid. Water containing earthy carbonates is usually bright, hard, pleasant to the taste, and refreshing, while pure water is soft, tasteless, and affords none of that sense of exhilaration which is always associated with 'the crystal spring.' The French *savans*, when inquiring after water for the supply of Paris, found that more conscripts are rejected in soft water districts, on account of imperfect development and stunted growth than in the hard; and they concluded that calcareous matter in water is essential to the formation of tissues. In these islands, it appears that the death-rate is influenced by the water supply, not only as to its sufficiency and the amount of organic matter suspended in it, but also as to its relative hardness. Glasgow and Manchester are supplied with soft waters, and have high death-rates; Birmingham, Bristol, Newcastle, and Warwick have hard waters and low death-rates. It may be said that in towns supplied with water of more than ten degrees of hardness, the average mortality is about 22 per 1,000, while in those supplied with softer water it is about 26 per 1,000."—*City Press.*

Body and Soul Emanations.—I have alluded to the mundane emanations, and we must not omit all mention of the emanations from the body, and their action upon those with whom we are brought into close contiguity. This topic, it might be considered, would come more properly under Mesmerism, but there is a wide field that Mesmerism does not include. The sun has an atmosphere, the world has an atmosphere, and so has man, and one often evident to the senses; but it is not this to which I allude. We all know how we are drawn to some people, and repelled by others, when we are brought within their sphere. But the character of the emanations surrounding us has never been very definitely determined even where its existence has been admitted. Of its existence, however, there can be no doubt, although it is not recognised by physiologists or medical men except in some cases of disease. In Australia there are large patches of the sensitive plant growing wild, which curls itself up at the *approach* of man, not waiting for his contact, and some people are equally sensitive to the approach of others. We are told by Humboldt that the Peruvian Indians in the darkest night can not merely perceive through their scent the approach of a stranger whilst yet far distant, but can say whether he is Indian, European, or Negro. This is attributed to smell, but of the general character of this influence less is known. Dr. Bird, however, says: "It appears that every substance, while moved by heat, in turn changes heat. The sun's heat, changed by everything it permeates, passes on into the tissues, and there becomes transmuted into motion or organised tissue, which is bottled heat of peculiar modification. This latter again becomes heat when it moves and does work. This released heat takes its character from the tissue which has released it. Heat from brain tissue is not the same as that given out by muscular tissue." Dr. Bird applies this to his own department of medicine, for "as different heats yield to the influence of different things, it follows that a different substance is necessary to control the heat issuing from the brain from what is necessary to control heat issuing from muscle." This "heat or force theory rests on the broad and sure basis of physical science in all its departments, and by means of it we are connected and interwoven, not only with the earth and everything that moves on it, but with the sun, the moon, and all the radiant stars."* It is this "released heat taking its character from the tissue which has released it" that determines the character of the special bodily atmosphere that surrounds us. In some exceptional cases large quantities of this specialised force—this released heat taking its character from the tissue—is given out, and we have transmitted muscular, electrical, vital, or nervous forces in great power. Of pure muscular force we have such cases as those of Angelique Cattin, investigated by M. Arago. Electrical force is sometimes in great excess. All chemical action is attended by a generation of electricity, and this is so largely developed in some people,—their bodies have been so highly charged,—that it has been necessary to carry it off by a regular process of conduction. Vital force is so strong in others that they possess a curative power by its transmission in cases where healthy vital force is deficient. Old people imbibe vital force when sleeping with young ones, and all persons deficient in vital force draw largely from those with whom they sleep, causing or greatly increasing rheumatic and other pains. In the transmission of force from brain tissue we not only transmit mental states, but these other forces come more or less under the dominion of the Will, accounting for much in Electro-biology, Mesmerism, and so-called Spiritual Phenomena. But these are abnormal states; in the natural state the "released heat taking its character from the tissue," mixed with more material emanations, forms our personal atmosphere, and people are much

* "Physiological Essays," "The Origin of Disease," pp. 245, 246.

more under its influence than they are disposed to credit. Our thoughts and feelings are greatly influenced by those with whom we come in contact, and especially by those with whom we habitually associate, the influence depending upon the particular brain tissue from which the force emanates. Envy, hatred, and malice transmit themselves, as do faith, hope, and charity. Good people sometimes wonder at the devilish thoughts that come into their minds, and trace them to evil spirits; and such is really their origin, the evil influence or *spirit* being generated in another person's brain. The extent to which we give or take depends upon the constitution. Highly nervous people are very sensitive to the impressions about them. "When I am among women," writes Keats, "I have evil thoughts, malice, spleen; I cannot speak or be silent; I am full of suspicions, and therefore listen to nothing; I am in a hurry to be gone."* He was probably more susceptible to female influence, as the temperament of women is more sensitive, and approaches nearer to the poetical; but the influence he would receive would depend entirely upon the character of the women he was with, and upon their predominating brain development.

Shelley was equally sensitive. Medwin, writing of him, says: "So sensitive was he of external impressions, so magnetic, that I have seen him, after threading the crowd in Lung' Arno Corsos, throw himself half fainting into a chair, overpowered by the atmosphere of evil passions, as he used to say, in that sensual and unintellectual crowd." These phenomena are now illustrated on a very large scale in what are called Spiritualist circles. All that is wanted is observation and experiment; but we must look in the right direction—for *spirit*, not *spirits*, and for nervous and bodily forces and emanations.

The effect of one body upon another, even without contact, is familiar to us in disease. Mad people, or the mentally

* Monckton Milnes's (Lord Houghton), Keats, p. 245.

diseased, are known even by their smell, and it is now very generally admitted that it is very undesirable to crowd certain diseases together in hospitals; that patients recover much more rapidly by separate and isolated treatment.* Mad Houses are, I should think, much more likely to make sane people mad than mad people sane. But we may equally *catch* health as disease. Mr. Valentine Greatrakes, in Charles the First's time, wrought great cures, and the Royal Society of that time, then in its infancy, accounted for them by "a sanative contagion in Mr. Greatrakes' body, which had an antipathy to some particular diseases and not to others." If the Royal Society had pursued this investigation we should not have had people, in possession of this power, mistaken for impostors, and nearly torn to pieces by the mob. Such an investigation would also have thrown light upon many important kindred subjects connected with our higher interests—for instance, the way in which mind acts upon mind. Of the cause of our elective affinities, our unaccountable bodily likes and antipathies, we know nothing, and to pursue the subject would be to trench on Electro-biology, Mesmerism, Spiritualism, &c. George Whitefield's extraordinary power as a preacher depended entirely upon his personal presence; his published sermons read very tame,

* Recent returns show that the percentages of death after limb amputations under different conditions and degrees of aggregation and isolation are as follows:—In the large Parisian hospitals, 62 in 100 die; in British hospitals, with 300 to 600 beds, 41; with 300 to 201 beds, 30; with 200 to 100 beds, 23; with 100 to 26 beds, 18; 25 beds or less, 14; in isolated rooms in county practice, 11. Sir James Simpson argues from this that the congregation of patients upon a given spot, or within a given establishment, enhances the chances of death to those that are there operated upon and treated, while on the other hand the more that patients are separated and isolated, the more surely they recover from the knife of the surgeon, and in all probability also from other accidents and disease. There is safety, he maintains, in segregation, danger in aggregation.—*(Pall Mall Gazette.)*

and create little impression. It was his power of throwing his mind into that of others, owing to his singleness of purpose and earnestness to which the effect was owing; as illustrated in an abnormal state in electro-biology. "The man was," as the *Athenæum* says (April 22, 1871), "transparently sincere and single-hearted. It was out of a soul at white heat of emotion that the dramatisation proceeded; and the emotion that fired and fused it was the love of souls."

T. Adolphus Trollope, speaking of the influence of physical love on one of his characters, says: "It really seemed as if there were some physical emanation from her person—some magnetic stream—some distillation from the nervous system of one organisation mysteriously potent over the nervous system of another, which mounted to the brain, mastered the sources of his volition, and drew him helpless after her, as helpless as the magnetized patient obeys the will of the magnetizer." *

Instinct.—Whatever principle of agency in the animal is not recognised as an element in the human constitution has hitherto been called Instinct. Instinct, however, is hereditarily transmitted experience. Man can transmit his experience by oral and written records, whereas animals can transmit theirs only in the form of instincts. Voluntary acts many times repeated become involuntary or automatic—that is, take place unconsciously. This is no doubt accompanied by change, or even growth of structure, and this alteration of structure is transmitted to offspring with its function of automatic action. In this way we get the performance of an act without knowledge or experience, but the knowledge or experience has been acquired in previous generations. Instinct is only memory once removed. Memory is the result of impressions on the brain; these impressions are deepened by repetition till both speech

* "A Syren," vol. 2, p. 224.

and action become involuntary in a recognised and definite order along the path so often travelled. In old age, when our animal vigour is exhausted, and less force passes through the brain, and the brain itself becomes less susceptible of impression, the old impressions resume their sway, and we return to our old habits of feeling and thinking, and our early memories. As we have said, when these automatic memories are attended with alteration or increase of structure, and are transmitted to offspring, they become instincts. Man is thus a bundle of instincts, transmitted through every form of life which has existed since it first commenced within the simple cell millions of years ago. All his Feelings and Intellectual Faculties are instincts, and have been thus transmitted. Even his Reason is an instinct—a comparing, fitting, and generalising instinct, dependent upon transmitted structure of brain in its three separate organs of Comparison, Causality, and Congruity. The nearest approach to man in bodily form, although not in the mental, is the ape; but the ape has to be traced down to its beginning in the monad. All man's instincts exist in some exaggerated form in some of the creatures below him, except, perhaps, a few of the highest, which have been developed only in him, but which nevertheless exist in a rudimentary form in other animals; for it is impossible to affirm that brutes have no Conscience, or sense of right and wrong; or that they have no Veneration, that is, respect or reverence; or no Benevolence; or that they do not Reason. The industrial and social polity of the bee is superior to man's, and the ant at least equals him in some of his vices, for one tribe goes to war with another for the sole purpose of making prisoners, which they treat as slaves, and upon whose labour they live idly. Horses and Dogs, though they have the germs of higher affections in them, show nothing like the governing instincts of ants; and even the highest apes, Mr. Darwin tells us, show nothing like the sense and love of beauty evinced by many tribes of birds.

Man came last into the world, and although he has all the mental powers that distinguish other animals "in one," yet he possesses some of them in a very inferior degree. His pride, however, disclaims any relationship to his fellow-creatures lower in the scale of being, so he makes a difference in kind instead of in degree between himself and them, and where he recognises this mental superiority, or indeed any kind of intelligence he does not understand, he calls it Instinct, which is a mere name to cover his ignorance, and to support a foregone conclusion about some kind of soul which the brutes are supposed not to possess. People having no knowledge of their own minds cannot be expected to understand those of animals. Our intimate union with the Dog, Cat, and Horse must convince us that their feelings in most cases are like our own, and that the intellectual powers that guide these feelings, of perception, conception, memory, and even judgment, or the adaptation of means to ends, are the same as our own, differing only in degree. Mr. A. R. Wallace, a most acute observer as a Naturalist, is of opinion that birds learn to sing in the same way as young ladies do—by taking lessons. He says facts "render it certain that the peculiar notes of birds are acquired by imitation, as surely as a child learns English or French, not by instinct, but by hearing the language spoken by its parents."* He is even of opinion (in which perhaps we cannot so readily agree with him) that birds build their nests by their own observation and experience, and not by instinct—that is, transmitted aptitude and experience. They must be aided at least by a large organ of what Phrenologists call Constructiveness, which is a building instinct. He says: "During the time they are learning to fly and return often to the nests, they must be

* "On Natural Selection," p. 222.

able to examine it inside and out in every detail, and as we have seen that their daily search for food invariably leads them among the materials of which it is constructed, and among places similar to that in which it is placed, is it so very wonderful that when they want one themselves they should make one like it?"* Very wonderful, we think, even supposing that they should have such precocious ideas of commencing housekeeping. But then Mr. Wallace tells us that when they really do begin to build they are assisted by the experience of the older birds; that young and old birds pair together, and that the young one consents "to be guided to some extent by its partner." The fact is, we do not understand the language of the creatures below us, the study of which would be much more profitable, and very much more interesting, than Greek and Latin; for we should then learn how Nature in its infinite experience has found out best to do most of the things we wish to accomplish, has stereotyped the knowledge, and handed it down to us in what we call Instincts.

The principle by which Instincts are formed is very simple; it is but the law of hereditary descent in its most permanent form. We begin with a creature all stomach, and the simple appetite, through the instrumentality of pleasure and pain, does all the rest. We have increased effort, which is increased exercise, leading, as we have seen, to increase of size; and with "growth exceeding a certain rate," with new conditions or outer relations, we have new organs and new functions. Certain habits are acquired, and the brain "grows to" the kind of activity thus become habitual, both in intellect and feeling, and the character thus impressed is transmitted to offspring. This condition of brain is often acquired unconsciously in what has been called "unconscious cerebration;" the brain grows in sleep and acquires strength and firmness; and processes of thought, originally

* "On Natural Selection," p. 222.

difficult, attain clearness and become easier. Our feelings, too, are unconsciously bent, and under new circumstances we often become vividly and sometimes painfully conscious of the new condition the brain has acquired—when the course of true love, for instance, has ceased to run smooth. In this way man began his education, not with the race, but with the first living creature, and much of the experience he possesses has not been acquired by himself, but by previous generations.

"The corollary here drawn from the general argument is that the human brain is an organised register of infinitely numerous experiences received during the evolution of life, or rather, during the evolution of that series of organisms through which the human organism has been reached. The effects of the most uniform and frequent of these experiences have been successively bequeathed, principal and interest; and have slowly amounted to that high intelligence which lies latent in the brain of the infant—which the infant in after life exercises and perhaps strengthens or further complicates—and which, with minute additions, it bequeathes to future generations. And thus it happens that the European inherits from twenty to thirty cubic inches more brain than the Papuan. Thus it happens that faculties, such as music, which scarcely exist in some inferior races, become congenital in superior ones. Thus it happens that out of savages unable to count up to the number of their fingers, and speaking a language containing only nouns and verbs, arise at length our Newtons and Shakespeares." *

Although Mind, then, is dependent upon the Brain, and power is generally proportioned to size, we yet see upon how many other things more or less perfect mental action is dependent. It depends upon the amount of Force we are able to extract from the Food we take, and the mode in which, through a more or less well-balanced structure or

* "The Principles of Psychology," p. 470, by Herbert Spencer.

bodily organisation, that force distributes itself through the system. The digestion, heart, lungs, liver, and other organs must be sound, and these bodily conditions are easily discernible to the practised observer in what are called the Temperaments and other appearances. Then there are Race, Family, Climate, Education, and Idiosyncracies, and Aptitudes, which circumstances and education give and transmit. By these mental aptitudes the children of educated parents can be distinguished from those of the uneducated. These aptitudes may be transmitted without increase of size; with increase of size they become instincts. An educated Phrenologist estimates and makes allowance for all these conditions, in predicating the power of the Mind from the organisation of the Brain, as shown in the shape of the head.

We shall proceed, then, to the consideration of the faculties of the Mind—its Propensities, Sentiments, and Intellect.

## The Propensities.

### THE SELF-PROTECTING.

Life is the most valuable thing we possess, and Nature has protected it not only by the pain which is always inflicted whenever it is endangered, but by an instinct making us wish to preserve it even in the midst of the acutest suffering. Long-continued pain may make us wish for death, but that very much depends upon the strength in which this "Love of Life" for its own sake exists, and this greatly varies in different individuals, and in different countries. Its sacredness increases with civilisation, as safety fosters the feeling. Life among savages is so precarious that it is little valued; and even among the Chinese a man will do another the little favour of having his head cut off for him as a substitute for about £25. It is true that this head may be comparatively of even less value, considering that 33 millions of people die annually; but we in this country are in the habit of considering that it would be a considerable loss to us. We learn also from a blue-book, "Correspondence Respecting Japan," lately presented to the Foreign Office by Mr. Mitford, that the Japanese in full Parliament refused to abolish the Seppuka, or Hara Kiri, which is suicide by disembowelling, because, as it was said, such suicide was "the very shrine of the Japanese national spirit, and the embodiment in practice of devotion to principle." Accordingly, when a Ministry is turned out, they make it a principle to "go out" altogether; they have their whitebait dinner, and then cut themselves open. We greatly admire "devotion to principle," but trust that it may not go quite to such extremes among ourselves. However desirable it may seem in some particular cases, we must not yield to mere expediency. It is this dread of annihilation, as it is called—this instinctive longing for continued existence—the object of which is the protection of life here, making us

Suffer
The slings and arrows of outrageous fortune,
Rather than take arms against a sea of troubles,
And, by opposing, end them,

which is adduced as conclusive evidence for continued existence beyond this life; for, it is said, why should we have had such a strong wish to live again, or to have existence continued, if there is to be no "Future State?" Thus Dr. J. H. Stirling says: "Then the dignity of the soul—its possession of faculties far above the needs of earthly commodity! All the natural desires possess their natural objects, and there is no desire more natural than the desire of immortality." The noble Roman thought death by his own hand preferable to dishonour, and the time may come when all will think "painless extinction" preferable, in the highest moral sense, to *cureless* suffering and to being a pain and burden to all around us. Seneca says: "The lot of man is happy, because no one continues wretched but by his own fault. If life pleases you, live. If not, you have the right to return whence you came." *

And, again, Julius Cæsar: "In affliction and misfortune death is the termination of our sufferings, and not a punishment; it takes away all the ills of humanity; beyond are neither cares nor joy."

But if a man wishes to live, he must eat to live—not, however, live to eat; and accordingly he has a desire for Food, and if this desire is not gratified he suffers the pain of Hunger. If he is not regularly "stoked," *i.e.*, supplied with the carbon, the union of which with the oxygen of the atmosphere supplies the force that works his machine, he soon comes to a dead stop. Consequently, although it is a man's first duty to take food, as he could not live if he did not, yet it is not left to his conscience. He does not take food because he thinks it right, but because Nature pinches him awfully in the stomach if he does not. When Conscience shall pinch as sharply for neglecting other duties,

* In one of the essays by members of the Birmingham Speculative Club (Williams and Norgate, 1870,) called "Euthanasia," the following proposition is enforced:—"That in all cases of hopeless and painful illness, it should be the recognised duty of the medical attendant, whenever so desired by the patient, to administer chloroform—or some other anæsthetic as may by-and-by supersede chloroform—so as to destroy consciousness at once, and put the sufferer to a quick and painless death; all needful precautions being adopted to prevent any possible abuse of such duty; and means being taken to establish, beyond the possibility of doubt or question, that the remedy was applied at the express wish of the patient." "Chloroform smooths our way into this world; why should it not smooth our way out?" says the *Saturday Review*. This question, I think, is now fully open for discussion.

these duties will be much more punctually performed, and little preaching will be necessary. But Food is not only requisite to the support of the body, but the necessity for taking it, and providing it, is the great support to our moral system. It is the foundation of all order and obedience—the rod which Nature holds over us *to oblige* us to do most of our duties. A man cannot be idle—he must work; it is of no use his rebelling and falling out of the social ranks; he must come home and take his place at feeding time. This necessity for taking food, and the hunger that attends it, are the great tamers of brutes, and of man no less.

But, although it is absolutely necessary that we should eat, and that several times daily, it has by no means been determined *what* we should eat. Experiments have been tried in Workhouses and Prisons without at present much practical result. We have dietaries and tables expressing food-values which place milk at the bottom of the scale and herring at the top! Liebig's food theories have been considerably shaken, and we have Letheby's Cantor Lectures "On Food and its Varieties" succeeding Liebig. He deals with its "chemical composition, nutritive value, comparative digestibility, physiological functions and uses, preparation, culinary treatment, preservation, adulteration," &c.; in fact, he is as exhaustive as he can possibly be in four lectures. Dr. Letheby starts first with wheat, and then comes the disputed question as to whether it is best eaten with the bran or without, and this can only be decided by the requirements and capabilities of each individual. The bran is considered to be rich in soluble phosphates, which are especially valuable in the dietary of growing children, and the bones and brain require them. But although useful to some on that particular account, it is injurious to others from the too great irritation of the alimentary canal it occasions. The hard-working navvy, the doctor tells us, chooses white bread because he finds it not only more digestible, but stronger, enabling him to do more work. As to Vegetarianism, or living altogether on vegetable food, perhaps if man had four stomachs like the ox, and the same love of, and time for, rumination, he might be able to extract enough "force" from a vegetable diet only; but as he is at present constituted there is but little work to be got out of him upon it. He wants a more nitrogenous diet, which is only found in flesh-food. In fact, if you want to take all the pluck out of a man you have only to keep him on vegetables. The experiment has been frequently tried both on individuals and communities. Meats are not to be judged altogether by their nutritious qualities; their digestibility must also be taken into account. It is of little use that food should contain a large amount of force, if the greater part of that force is consumed in digestion. Fish is considered to be of little value as food; but this may be prejudice, for not only does it contain some 18 per cent. of nitrogen, but Prof. Agassiz affirms that it supplies the best kind of food for the brain. Butter is required to supply carbon, and "all fish," we are told, "are in their best condition at the time of the ripening of the milt and roe; for not only are they fatter at that time, but when cooked they have a better flavour, and the flesh is solid and opaque. On the other hand, when they are out of condition, the flesh is semi-gelatinous and watery."

Gastronomy should become a science, for however much ashamed some well-bred young ladies may profess to be of eating and drinking, Gastronomy is of more importance to us than Astronomy, and by all the difference, at least, of their comparative near or distant relation to us. We know how much is done among both plants and animals by judicious feeding, and if the subject were properly studied the inherited defects of both bodily and mental constitution might be greatly diminished. No doubt it is in our power to make either queen bees or workers. A German philosopher says, "No phosphorus no thought," and certainly the degree of intelligence is largely in proportion to the amount of phosphorus in the brain, rising in quantity till we reach the madman, whose over-activity of brain constitutes his madness. Surely food might be judiciously applied in alleviation of such cases. In fact, we have no science on the subject, and, the medical profession notwithstanding, we are in a state of complete barbarism as to what to Eat, Drink, or Avoid.

Next in importance to taking Food, and next in the order of development, is self-defence. This instinct supplies courage for attack and defence; it gives pleasure in fighting, in aggression, and in meeting and overcoming difficulties. It often shows itself as mere Oppositiveness, as in the man who said that if he had any doubt about how to act he always asked his wife and did just the reverse.

Animals live upon one another, and thus the race is kept ever young, and their numbers within the means of subsistence; and in harmony with this necessity of their existence they have a pleasure in killing and destroying, independently of the mere gratification of appetite. Man, requiring flesh-food, is no exception to this law. It is illustrative of the barbarous age in which we still live, that, even among what are called the "upper classes," everywhere killing is still called "sport." Man never meets a fellow-creature of the orders below him, placed under his protection, to whom he is as God, than the first thought of his heart is murder—for to kill anything *unnecessarily* is murder. It is quite time this brute-beast part of our nature took a higher direction, and that our sympathy was with a creature's enjoyment in *its* world, and not in its destruction. Energy of character is the best form of this feeling, and its worst is envy, hatred, malice, and all uncharitableness. It gives a strong desire to give pain to those who give pain to us, and so far is self-protecting. The existence of this Propensity, its usefulness, and indeed the absolute *necessity* for its existence, is evidence that Love, in a world constituted as ours is, is not to have it all to itself. It is as desirable that we should hate ugliness—in all its forms of vice, error, untruthfulness—as that we should love that which is loveable. In fact, it is difficult, as things now are, to say which is the greater virtue—Love or Hate; for it is as necessary now as in St. Paul's time "to be angry and sin not"; and "hates any man the thing he would not kill?" is a right direction of feeling against evil.

The feeling of hatred and the desire to destroy the nearest relatives, as well as to love them, has been transmitted as an instinct, when of service to the community, and Darwin says, "if men were reared under precisely the same conditions as hive-bees, there can hardly be a doubt that our unmarried females would, like the worker-bees, think it a sacred duty to kill their brothers, and mothers would strive to kill their fertile daughters, and no one would think of interfering."* In fact, that it would be possible to work this world for the greatest good of all on the principle of love alone is only a fond delusion. You see a poetic-looking gentleman, for instance, greatly admiring the young lambs, "as they skip and bound as to the tabor's sound," and you think he is sympathising with Wordsworth. You find it is the "fore-quarter" only he is thinking of; and the sympathy of the lady when meeting the butcher was probably not much more genuine—"You are not going to kill that dear little lamb, are you, butcher?" "Why, ma'am," said he, "would you eat it alive?"

It has been in the struggle for existence—in *fighting* for dear life; in sexual selection, or in *fighting* for love—that mind has been developed and perfected. "How much Nature seems to love us!" says Nathaniel Hawthorne; and

* "Descent of Man," vol. 1, p. 73.

how readily, nevertheless, without a sigh or complaint, she converts us to a meaner purpose, when her highest one—that of conscious intellectual life and sensibility—has been ultimately baulked!"*

But an animal cannot always protect itself by open force, and it must make up in cunning—our old word for wisdom—what it wants in strength or courage; "it ekes out with the skin of the fox what is wanted of the lion's."

The use of this faculty is Concealment; its abuse, Evasion and Deceit. No doubt the feeling was slightly in excess in the man who, on being asked by the priest on his death bed if he repented, said, after due consideration, "No, he thought he had done as many as had done him."

An animal, if it would take proper care of itself, must not only appropriate what it requires for immediate consumption, but must lay by something "against a rainy day," and some animals, and all men, have a propensity to do so. This feeling is the base of the law of Property, and greatly aids accumulation; in abuse, it makes a man who devotes himself entirely to business, and a miser. There is also a mental acquisitiveness—a love of knowledge for its own sake, which in excess makes a book-worm. In the curiosity of monkeys, (of which Mr. Darwin gives many instances,) Sir Alex. Grant tells us there is the commencement of that love of knowledge for its own sake which is one of the noblest of attributes. Women often show a laudable curiosity in things which do not in the least concern them: doubtless a love of knowledge for its own sake, the *commencement* of a very noble attribute.

No doubt man and some other animals have a propensity to build both a home and a place for their store of savings—

* "The Blithedale Romance," p. 296.

a building instinct, which greatly aids, if it does not entirely direct, the bird, the bee, the beaver, &c. "Man is a tool-using animal, without tools he is nothing, with tools he is all," and this faculty gives him a pleasure and a power in their invention.

As Love and Hate are necessary—if not equally necessary, parts of our mental condition, so are Courage and Fear equally self-protecting. An instinct that makes us avoid danger is perhaps more necessary than that which induces us to meet and overcome it. A child with small cautiousness is always in danger. To the timid hare, with its large ears and long legs, it is its principal protection. In man it produces prudence and circumspection; and in excess it induces cowardice, indecision, night-mares, depression of spirits, and, like pain itself, lowers vitality.

## THE SELF-REGARDING FEELINGS.

Pride and Vanity represent these feelings in excess; in their due proportion they are Self-respect and Love of Approbation—personal dignity and the desire of the esteem and praise of others. Our estimate of ourselves is not often in accordance with our real worth, but dependent upon the instinctive promptings of our self-esteem, and it was a wise observation of Lavater, that whoever makes too much or too little of himself has a false measure for everything—a trifle, plus his ego, is immense; an immensity, minus his ego, is a trifle. Neither is our ambition always a laudable one. The painful self-consciousness resulting from the excess of these feelings is familiar to us all; but pride and vanity are the vices of the present age. All are seeking distinction in some form or other, and in ways not always the most ennobling, or much above the creatures that are considered so far below them. The peacock spreads its tail, and so does a lady, and both with the same object; but the peacock has this advantage, however, of having his tail already sufficiently ornamented, while the lady has to follow all the vagaries of fashion, and wear besides any kind of fools' cap fashion may require. Men, and especially women, are so fond of praise themselves, that they praise their gods in a way which, if addressed to one of themselves, would be considered fulsome in the extreme, particularly if its sincerity were at all questionable, and there was some object to be gained by it. In the earliest ages flocks and fruits were offered as a most acceptable sacrifice; they now offer what they themselves like better still, praise.

Love of Power and of the Approbation of our fellow-men are the leading springs of action in the present day, and too often to the exclusion of higher motives. Thank God, we have those amongst us who can appreciate those higher motives, and all, therefore, will learn to do so in time. W. R. Grove, lecturing to the students of St. Mary's Hospital (May, 1869,) said: "The road to success is often that which a high-minded man cannot travel: he cannot learn to fetch and carry, to subserve the interest of a patron or a mob. I do not seek to undervalue success: duty to yourselves and those whom you may bring into the world enjoin its pursuit in moderation. But I would fain endeavour to inculcate upon my younger hearers a higher motive than the mere hope of fame, wealth, or power. If these come by an unswerving career, make good use of them; if not, console yourselves with the conviction that those who are said to be in power are frequently the veriest slaves in existence." And, again, *our* Carlyle (we may be proud to call him ours): "Power! Love of Power? Does 'power' mean the faculty of giving places, of having newspaper paragraphs, of being waited on by sycophants? To ride in gilt coaches, escorted by the flunkeyisms and most sweet voices,—I assure you it is not the heaven of all, but only of many! Some born Kings, I myself have known, of stout natural limbs, who, in shoes of moderately good fit, found quiet *walking* handier; and crowned themselves almost too sufficiently, by putting on their own private hat, with some spoken or speechless, 'God enable me to be King of what lies under this! for Eternities lie under it, and Infinitudes,—and Heaven also, and Hell. And it is big as the Universe, this Kingdom; and I am to conquer it, or be for ever conquered by it, now while it is called To-day!'" *

On this much-craved approbation of our fellows, Jean Paul F. Richter said, a hundred years ago: "On the whole, I hold the constant regard we pay, in all our actions, to the judgment of others, as the poison of our peace, our reason, and our virtue. At this slave-chain I have long filed, and I scarcely ever hope to break it entirely asunder. I wish to accustom myself to the censure of others, and *appear* a fool, that I may learn to endure fools." We are told that "three years after his marriage Herder was invited to Weimar, to fill the place of Consistorial Rath and Court Preacher. Many reports had preceded Herder of his heresy and contempt of forms. They had said, among other things, that he preached in boots and spurs, and that after every sermon he rode three times round the church and out of the door on horseback." † There is nothing too ridiculous or too vile for people to say, and yet the world continues to put this talk in the place of conscience, and even makes a conscience of it, and a rule of right and wrong!

## THE SOCIAL AFFECTIONS.

The Social or Domestic Feelings include the Love of Wife, Children, Friends, and Home.

The love between the sexes is the strongest feeling in our nature, as it is also the second in importance; the continua-

* "Cromwell's Letters and Speeches," p. 228.

† Carlyle's "Life of Richter," appendix, p. 214.

tion of the species, and the world that is to follow us, depending upon it. It has a whole brain to itself to preside over its functions; and as man has a brain for himself—the Cerebrum, so he has one for his offspring, or the world that is to be—the Cerebellum. It is necessary, however, having regard to the functions of both brains, that there should be one man and one woman to make a complete human being; and the highest interests of humanity require that our efforts should be directed towards their completeness of union. If this be true, then the relations between the sexes, so much now in dispute, must be settled in accordance with it, and the kind of education and condition required is that which will most harmoniously unite two people together *for life*. Now, what does a man want for his other half? Not another man, or anything like one, but a true woman, feminine in all her attributes, modest, clinging, confiding, caressing, affectionate, trusting, moved by her highest instincts, which in her are above reason. And what does a woman most want to secure her faith and fidelity? Why, a true man; manly, strong, independent, courageous, upright, truthful, magnanimous. To the one belongs power, to the other beauty; one has the largest brain and most strength of mind, the other is the most sensitive, and can see and feel what the man cannot; it is the province of one to subdue and to rule the world, and of the other to beautify it; in the woman the social feelings, which attach her to husband, children, and friends, predominate; in the man, the moral feelings—faith, hope, charity, benevolence, justice, ideality. The woman's head is consequently long, the man's round and high;* each first and

* Aristotle's enumeration of the differences between man and woman does not indicate a very high opinion of the latter, but we may hope that she has improved since then. "Woman," he says, "is more pitiful than man, and more given to tears, and more prone to envy, and more querulous against fate, and sharper with her tongue and readier with her hands. The female also is more easily dispirited and less hopeful than the male, and less sensible of shame and less

highest in his or her department. The movement in favour of "woman's rights" and the "equality of the sexes" is taking a wrong direction. Now the sexes are equal—each first and best in their separate departments; but a woman attempting to compete with man in *his* department loses that on which her position principally depends, and at best becomes but a weak imitation of the man. A woman donning the "unmentionables" loses quite as much as a man who puts himself into petticoats.* It has become rather fashion-

careful of truth. * * The male is more ready to help and more courageous than the female; for even among the mollusca, when the cuttle-fish is struck with the spear the male helps the female, but when the male is struck the female makes off."

* Dr. Thomas Laycock is of opinion that his method and principles of psychological investigation help to illucidate the fundamental laws which govern the social relations of the sexes, and to indicate "how far the organology of woman determines the education proper to her." He has, therefore, in his 2nd edition of "Mind and Brain," added a short chapter on manhood and womanhood. He is clearly of opinion that, "strong in her maternal and moral instincts, woman will be able to do more for man's spiritual and moral elevation as a wife and mother than in any virile occupation whatever, however honourably she may fill it." The vigorous agitation in the United States for equality of women with men in education, trades, profession, and political power, he considers to be a proof of the fact "that a fundamental feminine instinct is enfeebled."

"The advocates of 'woman's rights' will do well to study the progress of Miss Shedden's case in the House of Lords. One can conceive how terribly the Lord Chancellor must have suffered before in answer to Miss Shedden's request to 'Allow her, for one moment,' he replied in these awful terms, 'No, not for one moment; this has gone on too long. Tell us which point you will go upon. Answer my question. You are trifling most improperly with the court. It is beyond all endurance.' Their other lordships concurred, no doubt with tears in their noble eyes. Even Mr. Bright will not utterly condemn this ebullition of feeling when he remembers that they have now been engaged nineteen days in hearing a case which one of the viler sex would probably have disposed of in as many hours. We cannot read the account without deeply sympathising with the lords and the lady. The proceedings have been interrupted by hysterics, and both sexes

able lately to question these principles. There have been so many ill-assorted marriages that the institution itself has been called in question, and people begin to think that a wife should be taken on the same terms as a house—on lease for 7, 14, or 21 years; but the fault lies not in the institution, but in the ignorance and utter disregard of the laws upon which two people ought ever to come together. The laws of physiology and psychology—that is, the best interests of body and mind—point to marriage for life. The sexual feeling is the greatest consumer of force, and when indulged in too early deprives both body and mind of their proper growth. It should never be allowed to act separately for the sake of its own gratification, as it does in the brutes; and it is only in marriage that it can be put under due restriction, surveillance, and control.

People fall in love often from no higher feeling than mere animal affinity. It is this first spark that requires carefully watching, that it may be stamped out at once if reason does not approve, for if it is allowed to attain any head it is very difficult to extinguish it, and reason is led captive. This love between the sexes, the strongest passion we have, is intended, in its early power, to smooth down individualities and blend two people together for life—two lives merged into one personality, as afterwards in the offspring.

In marriage this feeling unites itself first with the other

are utterly worn out. At the same time it cannot be denied that many of us in private life would not be sorry occasionally to call in the Lord Chancellor. The words 'Tell us which point you will go upon,' will, we fear, although wrung from the bitterness of his lordship's heart, find an echo in many a manly bosom. No woman ever yet answered the question; perhaps because they have not made up their minds; but we venture to express a hope that it is not their intention to go upon the point of arguing in public as they argue in private. It would be a cruel thing to take from men who suffer much at home the rest and relaxation they now find in hours of business." (*Pall Mall Gazette*, June 20, 1869.)

social feelings, and then with all our highest feelings, sharing its intensity of force among them, and ultimately they act and can act only together. The mere physical love is weaker, and put under due control; but married love is much stronger, because it comprises all our best feelings. This is the highest state, in which all our joys are doubled—all our sorrows halved; but it cannot be attained if people come together for a term, and marriage is dissoluble. "How many sins are condoned, how many shortcomings overlooked, when a man certainly belongs to a woman, who can tell? But suppose the relationship different; let him be merely her lover, whether sinfully or sinlessly, the moment the glamour with which, it may be, she has herself surrounded him is dispelled, she sees one fault after another, and dispassionately weighing him in the balance, finds him wanting." *

Of course allowance must be made for cases where such growth does not, and from the nature of things cannot, take place; but for people to allow themselves to wander about, seeking what they consider their natural affinities, and which are generally merely animal, is to give Amativeness undue prominence, and thus disturb the whole mental balance, and also to deprive themselves of that blessedness which is alone derivable from the complete union of two natures. Even if we were to consider the sexual feeling alone, its highest gratification, and the only lasting one, is when it is allied with all that is modest, pure, and good.† "It has been esta-

* "Austin Friars," by the author of "George Geith, of Fen Court."

† The history of this woman question is the history of the advancement of the race and of civilisation. Mr. Darwin's book on "Sexual Selection" certainly does not tend to raise our ideas of the sex, or indeed of human nature generally; and he tells us, no doubt very truly, that "he who has seen the savage in his native land will not feel much shame if forced to admit that the blood of some more humble creature flows in his veins.—(Vol. 2, p. 404).

Birds have been much longer in the world than man, and marriage is decidedly an institution among them. Mr. Darwin says: "The

blished from an enormous body of statistics, that the unmarried die in a much larger proportion than the married."—(Darwin's "Descent of Man," vol. 1, p. 176.)

females are most excited by, or prefer pairing with, the more ornamented males, or those which are the best songsters, or play the best antics; but it is obviously probable, as has been actually observed in some cases, that they would at the same time prefer the more vigorous and lively males" (vol. 1, p. 262); and we are told also that pea-hens, if they cannot have the male they have set their hearts upon, very sensibly determine to remain single.

Women make their selections on precisely the same principles. It is the males that "are most ornamented" and that play "the best antics" that are chosen; and we are all familiar with the story of the servant girl who spent all her savings in buying a soldier out of the army, but with whom, when he had doffed the red and donned his old smock-frock, she refused to have anything to do. We must not suppose that it is birds and men only who try to make themselves agreeable to the ladies of their choice. Frogs and toads, and tortoises, and even crocodiles and alligators sing sweetly when they are in love. "Frogs and toads possess vocal organs, which are incessantly used during the breeding season. * * The male alone of the tortoise utters a noise, and this only during the season of love. Male alligators roar or bellow during the same season."—(Darwin, vol. 2, p. 331). We may call it noise and bellowing, but no doubt it is sweet music to the lady-love, who also has her own ideas of the beautiful as well as of music. It is the same with man; as Darwin says, "It is certainly not true that there is in the mind of man any universal standard of beauty with respect to the human body."—(Vol. 2, p. 353). "The same fashions," he says, "in modifying the shape of the head, in ornamenting the hair, in painting, tattooing, perforating the nose or ears, or removing or filing the teeth, &c., now prevail, and have long prevailed, in the most distant quarters of the world." (p. 343.)

Hearne, who lived many years with the American Indians, and who was an excellent observer, says, in speaking of the women, "Ask a Northern Indian what is beauty, and he will answer, a broad flat face, small eyes, high cheek bones, three or four black lines across each cheek, a low forehead, a large broad chin, a clumsy hook nose, a tawny hide, and breasts hanging down to the belt" (Darwin, p. 344.) The Chinese think Europeans hideous with their white skin and prominent noses (p. 445.) Judged by the Mogul race, they have been described as having the beak of a bird with the body of a man.

The love of children is a strong instinct in human nature, and it is well that it does not depend for its exercise upon

"Man admires and often tries to exaggerate whatever characters nature may have given him."—(Humboldt.)

In our earliest history marriage was certainly not the law in the human race, and, as Sir. John Lubbock tells us, "according to old ideas a man had no right to appropriate to himself that which belonged to the whole tribe.—("Origin of Civilization," p. 86.) Sir John further gives a most curious body of facts showing that in old times high honour was bestowed on women that were utterly licentous; and this, as he explains, "is intelligible, if we admit that promiscuous intercourse was the aboriginal and therefore long-revered custom of the tribe.—(Ibid, p. 361.)

Woman, even so late as the beginning of the last century, was treated as a devil and burnt as a witch. She is now, in the Religion of Humanity, worshipped as a God. "To comprehend a witch-mania," says Canon Kingsley, "you must look at it as (what the witch literature confesses it unblushingly to be) man's dread of Nature excited to its highest form, as dread of woman. She is to the barbarous man—she should be more and more to the civilised man—not only the most beautiful and precious, but the most wonderful and mysterious of all natural objects, if it be only as the author of his physical being. She is to the savage a miracle to be alternately adored and dreaded. He dreads her more delicate nervous organisation, which often takes shapes to him demoniacle and miraculous; her quicker instincts, her readier wit, which seem to him to have in them somewhat prophetic and superhuman; which entangle him as in an invisible net, and rule him against his will. He dreads her very tongue, more crushing than his heaviest club, more keen than his poisoned arrows. He dreads those habits of secrecy and falsehood, the weapons of the weak, to which savage and degraded woman always has recourse. He dreads the very medicinal skill which she has learnt to exercise, as nurse, comforter and slave. He dreads those secret ceremonies, those mysterious initiations which no man may witness, which he has permitted to her in all ages, in so many—if not all—barbarous and semi-barbarous races, whether Negro, American, Syrian, Greek, or Roman, as a homage to the mysterious importance of her who brings him into the world. If she turn against him—she, with all her unknown powers, she who is the sharer of his deepest secrets, who prepares his very food day by day—what harm can she not, may she not do? And that she has good reason to

either our conscience or our reason, or the little brats would too often be laid aside and forgotten, like their prototypes

turn against him, he knows too well. What deliverance is there from this mysterious house-fiend, save brute force? Terror, torture, murder, must be the order of the day. Woman must be crushed, at all price, by the blind fear of the man."—(*Fraser*, June, 1866, p. 715.)

By the Positivists, in the "Religion of Humanity," woman is worshiped as the highest symbol of humanity. "As the symbol of humanity," says its High Priest in this country, "we adopt with somewhat altered associations the beautiful creation of the mediæval mind—the woman with the child in her arms; and to give life and vividness to this symbol, and to our worship in general, each Positivist adopts as objects of his adoration his mother, his wife, his daughter, allowing the principal place to the mother, but blending the three into one compound influence—representing to him Humanity in its past, its present, and its future."

M. Auguste Comte, the founder of this religion, had the merit, not uncommon, of teaching on this subject both by precept and example—by precept what we ought to do, by example what we ought not. In precept he recognised "all the possibilities of true marriage"; in practice he made, in 1825, "a marriage of convenience," which proved however to be no convenience to either of them, as 17 years after their marriage Mme. Comte left her husband never to return. It does not appear to have been a happy marriage: M. Comte did not at all worship his wife, but seems to have expected her to worship him. We are told that he was sent to school at the age of nine years, and was so precocious that at ten "he criticised with severity and judgment his teachers and their methods of instruction." Now he may have carried his admirable talent for criticism into private life, and his wife may not have appreciated it. We are told also that "she could not forgive in him the unconscious egotisms of a powerful genius." It was reserved therfore for another man's wife to develope the Religion of Humanity. Mme. Clotilde de Vaux had been separated from her husband, who had unfortunately been guilty of a capital crime, and sent to the galleys; but she became to Comte "a revelation of the power, purity, genius, and suffering of woman—so that, having worked out his theory of Divine Humanity, he recognised its highest development in her noble self-sacrificing life." Comte himself says, in reference to her; "Through her I have at length become for Humanity, in the strictest sense, a twofold organ, as may any one who has reaped the full advantages of woman's influ-

the wooden dolls. A woman considers it a great virtue to be fond of her children, but it is one she shares with the tiger, and a monkey is the fondest of its offspring of all animals.

"So intense is the grief of female monkeys for the loss of their young, that it invariably caused the death of certain kinds kept under confinement by Brehm, in N. America. Orphan-monkeys were always adopted and carefully guarded by the other monkeys, both males and females. One female baboon had so capacious a heart that she not only adopted young monkeys of other species, but stole young dogs and cats, which she continually carried about. An adopted kitten scratched the above-mentioned baboon, which certainly had a fine intellect, for she was much astonished at being scratched, and immediately examined the kitten's feet, and without more ado bit off the claws." *

Both Darwin and Sir John Lubbock place the Social among the Moral Feelings. Thus Sir John says, "We do

ence. My career had been that of Aristotle, I should have wanted energy for that of St. Paul, but for her. I had extracted sound philosophy from real science; I was enabled by her to found on the basis of that philosophy the universal religion." We perhaps ought not to feel surprised that his wife, sensible, practical woman as she is represented, could not forgive in him the unconscious egotisms of a powerful genius. There is considerable sympathy with this article in Comte's creed in the New World, although it is variously expressed. The Rev. John H. Noyes, a great authority on the subject of Spiritual wives, says "that when a man becomes conscious that his soul is saved the first thing that he sets about is to find his Paradise and his Eve." "It is a very sad fact, however," says Hepworth Dixon, "which shows in what darkness a man may grope and pine in this wicked world, that when these Perfect Saints were able to look about them in the new freedom of Gospel light, hardly one of the leading men among them could find an Eden at home, an Eve in his lawful wife."—("Spiritual Wives," Vol. 2, p. 10.) So men of genius, as a rule, have no objection to take a wife, but it is generally one belonging to somebody else.

* C. Darwin, vol. 1, p. 41.

not generally attribute moral feelings to quadrupeds and birds, yet there is perhaps no stronger feeling than that of the mother for her offspring."* We rather revolt against the "moral sense" of the tiger and the monkey. The moral feelings are rather those that unite man with his fellow-man, which turn man into humanity, upon which unity the superior strength of the race depends. The predominance of the Social feelings would confine our sympathies to one family and clan, whereas we belong to the world.

Attachment to individuals, which when fully developed in the character creates a warm and affectionate disposition, is the foundation of Friendship and of Society, and is as often misplaced as the love between the sexes. People leave cards and dine with one another, keeping the balance of such visiting tolerably even, and a man boasts of his large circle of friends. When he becomes poor, of course the *friendship* ceases with the dinners, and to complain then that his friends have deserted him would be absurd. "Love, and love only, is the loan for love," and dinners form a very imperfect substitute. Friendship, to be true and lasting, like love between the sexes, must be based upon the higher sentiments. Women, however, in whom the instinct is very strong, often prove an exception, and like a person the more for ill-treating them: as Moore the poet says, "They love on thro' all ills, and love on till they die."

The Love of Home is an important element in the Social Group, and as an instinct, among animals, helps to keep them to the places and districts to which they are best fitted by their nature. Among human beings it gives attachment to country and place, and indisposition to change of abode.

* "Origin of Civilisation," p. 265.

## THE MORAL FEELINGS.

The feelings we have considered have relation to our own individual well-being, and that of our family and friends; what are called the moral feelings induce us to seek also the welfare of our fellow-creatures—of our own race, and of all capable of enjoyment. Man erroneously looks upon himself as an individual, when he is simply a part and parcel of the forces of Nature around him. By the force of gravitation he is connected with every other atom in the universe; the sun prepares his daily food, and supplies the force that works both his bodily and mental machine; and if he were cut off from either his food or the atmosphere around him he would soon come to a dead stop. His moral feelings connect him with his fellow-creatures, and make him a part of the great body of Humanity. He cannot breathe properly away from his fellow-creatures, and a Robinson Crusoe is a fish out of water. The mind requires its food the same as the stomach, and every faculty must have its natural external stimulant.

A Sense of Justice, a disposition to do what is commonly called "right between man and man," is a strong feeling in human nature. Other animals respect the rights of their fellows from fear of the consequences; man from the offence it gives to this sense—the pains of conscience. Of course there are some in whom this feeling is very weak, and others in whom it is merely rudimentary, and such have little or no conscience. They are as morally blind as they would be physically without their eyes, and they have to make up for this deficiency with other feelings or senses. This feeling is a sort of moral scales, by which we weight what is due to ourselves and others. Action and reaction are necessarily equal in the physical world, and ought to be so in the moral; and this faculty produces a sense of injury and injustice if the scales are not kept even, and we do not get as much as we give. It is all very well to inculcate disinterestedness and generosity, but if we were to give only and receive nothing we should soon have nothing to give; accordingly the world recognises that we should "be just before we are generous,"—just to ourselves as well as to others. At present we have a very low standard for the rights of others; man tranfers his accountability in that respect from this world to the next, like the Irishman who wished to restore the pig he had stolen when it appeared in evidence against him on the judgment day, as the priest told him it certainly would do. There are some very delicate consciences. I knew a little girl who would not throw the snails into a neighbour's garden, but put them on the top of the wall with their heads that way. The King of Burmah, a great theologian, explained to Dr. Bastian that Buddhism forbids the taking of life, that it is very wrong indeed to kill animals for food, and therefore you must always get someone else to do it for you. A pungzi, or, as we call him, a bonze, must not eat after noon; therefore he sits with his back to the sun, and being hungry at two or three o'clock, inquires of his scholar whether it is noon yet, to which the boy is trained to answer that it is not; and, on the strength of the pupil's assertion, the master eats another meal.* But we need not go to Burmah for illustrations of the supremacy of conscience: there are plenty of our own Clergy who knowingly turn their backs upon "the time of day" that they may continue to eat a comfortable meal at the expense of the Church. All the world over consciences are easily satisfied, resting in a few outside conventionalities; and the knowledge of what *is* right is generally behind even the disposition to do it. There is little justice in the world at present, or there would not be the great inequality in wealth that now prevails. The scales have never been held evenly between the different classes of society. A rich man takes a pound too much from a poor man in the division

* *Anthropological Review*, April, 1867, pp. 190, 191.

of the fruits of their joint labour, and returns him sixpence in charity, telling him not to make a beast of himself with it. Justice would make a much better ruler of the world than Generosity or Benevolence. To do what is right includes all the higher feelings; but all the feelings, high and low, must be put under the rule of this sense of justice, for that only can be right which is just. Carlyle says truly, "The just thing in the long run is the strong thing." We only want the "long run."

We are also told to "Love our neighbour as ourselves." Now as we cannot love from a mere sense of duty, but can love only that which is loveable, if our neighbour is not loveable we cannot love him. As, however, we have a natural desire for the happiness of others, a sympathy even with the enjoyment of all animated beings, we can try to make him happy without loving him very much. But if we really love to make our fellow-creatures happy, there is nothing *unselfish* in our doing it; we are only gratifying our own desire, and our neighbours therefore need not be too thankful, but gratify *their* desire, and do as they are done by. We are told that "the most wonderful and noble of all forces in the universe is disinterested affection;" and this is the kind of stuff goody people continue to talk. A *dis*interested affection, or one in which we are not interested, is a contradiction, for it would not be an affection at all. Such a *want* of feeling could never be a motive to action, for there would be nothing to move us. There is a love that induces us to seek our own well-being, and a love that induces us to seek the welfare of others; and it is the latter that is rather foolishly called disinterested, for if we have an affection for our fellow-creatures that induces us to seek their happiness in preference to our own, it is our affection or love that seeks gratification, not theirs; and in what way is that disinterested? In fact, we can only feel our own feelings, and if we do not possess the organisation upon which those feelings depend, we do not feel, and we do not act. The sooner, therefore, we recognise this fact, and leave off talking about "disinterested affection," and set ourselves to work to grow the organisation which *interests* us in the happiness of others, the better for the progress of morality.

> All who *joy would win*
> Must share it,—Happiness was born a twin.—*Byron.*

But if a man cannot love his neighbour—and he certainly cannot love by willing it—and if this desire for the happiness of others is not naturally a strong feeling with him, then he acts from a sense of duty, and does it by proxy or by deputy; and that is why benevolence, as it is miscalled, or charity, is probably doing more harm than good at the present time.* Or a man may act from a religious motive, and believing that he "that giveth to the poor lendeth to the Lord," trusts in the Lord, expecting to be paid *with interest* in another world. I fear it must also be allowed that if we have happiness in the welfare of our neighbours, we have also *some* happiness in their misfortunes, if we may judge from the lively and agreeable interest which their recital often awakens in us. The Elect, too, it is to be

* The world is full of Charities, left by our well-intentioned ancestors, that are sapping the foundation of all self-reliance and independence in their recipients, and pauperising whole districts in which they abound; but so fond are the trustees of these Charities of promoting the happiness of their fellow-creatures at some other person's expense, that they have resisted all efforts to divert these large funds to a useful purpose.

We have "Ladies' Charities" for bringing children into the world gratuitously, and finding all necessaries, although the parents have months to make such preparations for themselves; and we have School Charities, and Hospital Tickets, and Doles varying from 5s. to £4; and the poor get to think that there is no necessity for saving, although such independent providence is the foundation of most other virtues in them; and they spend more time in running after these charities and begging for them than they could earn double the amount in; and after all not one-third of those who have so misspent

feared, would not enjoy their heaven half so much if it were not for the lake of fire they know never can be quenched, but from which they are, happily for themselves, excluded.

their time get served, and the deficiency has too often to be made up out of the poor rates; the loss of character, however, nothing can restore.

"But that actual distress should be known to exist and the gifts of charity be withheld seems to most Christians a violation of the precepts of Christ, as well as a quenching of natural kindness. It must be right, they think, to give to him that needeth, and they will leave the consequences to God.

"Another habit of the same category is that of marrying early and in trust. * * Economic science, by studying the facts which come within its scope, and tracing effects to causes, has arrived at decided conclusions on these points. Under its teaching we know now many of which the best men were formerly ignorant. We now see carelessness is directly and inevitably produced by the chance of obtaining alms easily in time of need; and carelessness is the mother of idleness and sensual indulgence as well as of destitution. Benevolence on the part of the rich may *create* what the French expressively call *la misère;* it has no power to remove it. Where there is a hard struggle for the means of living, to marry and multiply without thought of the future is the way to keep down to the lowest point the condition of the whole labouring class. The accumulation of capital by saving is the only means of providing employment; and he who makes a fortune and invests it does much more for the poor than he who gives away all that he receives to the neediest people about him. The science that establishes these conclusions points, as a matter of course, to certain rules of conduct. If you wish well to the mass of mankind, you will endeavour to check waste, to increase production, to encourage industry and forethought and self-restraint. You will be extremely cautious not to put temptations in the way of the poor, by which—weak as they are by nature and circumstance—they may be reduced into thriftlessness. You will throw yourself heartily into the industrial efforts by which the fabric of material prosperity is built up."—(Rev. J. Llewelyn Davis, *Contemporary Review*, Jan., 1871, p. 197.) "In their readiness to minister without forethought to the bodily wants of their neighbours, Christians have really neglected the care of their souls. They have encouraged a dependence upon alms and rates, in place of a dependence upon the Providence of God."—(Ibid, p. 198.)

Veneration or Reverence is a strong sentiment in human nature, but not nearly so strong as it ought to be. The little knowledge that is a dangerous thing puffs up the present generation with a self-conceit proportioned to its superficiality, and "fools rush in where angels fear to tread." The feeling of Veneration begets courtesy and good manners, and respect for whatever is considered to be great and good. But men find superiority and greatness in wealth, not worth; and there is more worship of the golden calf now than in the most idolatrous times of the Israelites. From *ignorance* finding little worthy of veneration here below, man transfers his adoration and aspiration to another world, and not recognising God in his works, he makes himself a god after his own image, with his own passions and mental attributes, and the same love of power and vulgar taste for sitting upon thrones, and being served and obeyed, and praised and worshipped. A good man ought to be ashamed to be the deity that most people have thus made for themselves to worship.

"Hope springs eternal in the human breast," and, in proportion to the strength of the emotion, makes sunshine in the mind; and we have present happiness in the prospect of that which is to come. "It is like the sun, which, as we journey towards it, casts the shadow of our burden behind us." It is the counterpoise to cautiousness and fear.

The World, both intellectual and moral, as we conceive it, is a great delusion, "where nothing is, but all things seem;" but we nevertheless instinctively believe in its existence. Thus we have a nervous sensibility which we call colour, and we imagine it exists without ourselves, and not merely within the mind, and we cover the whole world with it; it is the same with the forms and sizes and localities to which we attach this colour: they are all created in the mind, and have nothing corresponding to them outside of us, and yet we have perfect faith in their objective existence. The action of all our faculties is attended with this instinctive belief—and yet there is nothing outside of us but one simple force, in various modes of action, acting upon the brain. We have equal faith in our own dreams, where there is no external stimulant calling the mind into action. Sometimes the brain action is reversed, faith, or the instinct of belief, calling into activity the intellectual faculties, and thus creating ghosts and phantoms, in which the belief is quite as strong as if the action were from without, from pure objective existence. We have equal faith in the action of the feelings. We have a sense of freedom and free-will, and we believe in it, although each force, and the action dependent upon it, is necessarily dependent upon some antecedent force. We have a sense of the Identity of the body, and we believe in it, although the body is never the same, but in constant flux with all around: the same of the mind, we have no evidence of its existence at all, except in its changes. The fact is, we may call these delusions, but the result of the action of each faculty is *truth to us*, and beyond that, *i.e.*, of absolute truth, we can know nothing. All that is intended, and that can be wished, is to produce modes of action which shall result in the largest amount of enjoyment. The airs that man gives himself about his knowledge are simply ridiculous, when all it can amount to is to enable him elaborately and laboriously to describe how the various modes of action without himself act upon his organisation.

The World which our faculties have created, the blue sky, the arch of heaven, the hills, the dales, the trees, the flowers with all their various gorgeous colouring, give us pleasure, which, connecting itself with organisation, has been strengthened in its transmission from generation to generation, ultimately resulting in a Sense of the Beautiful, a faculty which ought particularly to distinguish the Poet. This feeling refines the character, and begets the wish for more and more beauty—for higher and higher perfection. It is an exotic plant, and depends much upon cultivation, and it should be cultivated for the peculiar benefit of old age—"the soul's calm sunshine," when the passions have spent their force and are laid to sleep. The world, should appear more and more wonderful and beautiful to us as we grow older, as knowledge removes the veil, and as we recognise "the varied God."

### FACULTIES WHICH MAY EQUALLY AID ALL THE FEELINGS ACCORDING AS THEY ARE ALLIED.

There is a Feeling that gives a pleasure in retaining, and therefore a desire to retain, whatever feelings and ideas may at any time be uppermost in the mind. It is one of the most important feelings we have, as it gives a power of Attention, of concentrating the mind one on object—and of forming Associations and Habits. A small active brain will often excel a larger one without the power that this desire for permanence of emotions and ideas gives.

Firmness is another provision which nature has made for permanence. Evolution and growth, Nature's mode of progress, are slow, and we have feelings in harmony. Power of will depends upon this feeling, and we should learn to discriminate between firmness and obstinacy; the former being a strong will, the latter only a strong won't—the one a living force, the other a dead weight.

There is an instinct that induces us to copy the ways of the people around us; it helps to make the various members of the great body of society move harmoniously; the difference it alone makes between an Englishman and a Chinaman is very marked. It helps to make a good mimic.

There are feelings, as it seems to me, that have relation to the world within and to the world without our own minds. Thus there is a desire for permanence of emotions and ideas; there is also a desire for permanence of place, or love of home. There is a desire for the acquisition of property, and we have also a desire for the acquisition of knowledge; in excess the one making the miser, and the other the book-worm. We have a faculty that gives us pleasure in imitating the bodily peculiarities and actions of others; we have also one that reads and appreciates the natural language and characteristics of their different mental states, giving an intuitive power, with other faculties, of delineating character, as illustrated in Shakespeare and Scott.

The feeling of Personal Identity, or the "I" of consciousness, is also an instinct dependent upon the action of the brain, although phrenologists are not agreed as to its localisation, or even as to whether it is a primitive faculty at all. George Combe says: "Looking at the facts presented to us by observation, the most obvious inference seems to be, that the mind consists of an aggregate of powers, and that one of them supplies the feeling of Personal Identity, or the 'I' of consciousness, * * and it is extremely difficult to determine whether the feeling of personal identity indicated by the pronoun I is connected with a particular organ, or the result of the general action of the whole organs."* On the other hand, Mr. H. G. Atkinson says: "The I which represents the individualism seems to arise from a faculty whose organ is situated near to the Conscious sense and the Will, and close beneath Self-esteem and Firmness; 'I think; therefore I am,' is a conclusion from the Conscious sense and the sense of Personality." † The organ is liable to derangement and disease, when the feeling of identity is lost or

* "A System of Phrenology," 5th edit. vol. 1, page 172; vol. 2, p. 247.

† "Man's Nature and Development," p. 274.

double, and "one false case of identity takes away all reliance or arguments, as to the continuity or unity of being from the ordinary sense of identity."* Mr. Atkinson also gives a special organ to the Will, which he places next to the "I," under Firmness, and, what is quite as important, near Concentrativeness. As I have said elsewhere, "The Will is the *trigger* of the mind. Upon the 'last dictate of the understanding' it lets off the accumulated mental force in the direction of the object or purpose aimed at. The strength of the Will is in proportion to the force to be discharged, which force depends upon the size and quality of the organs in which the determination originates. * * The will no more lets itself off than the trigger of a gun does: it is generally pulled by the stronger motive, as understanding or mere feeling may prevail." †

The above is a tolerably complete list of the instincts or intuitive feelings at present known to us. No doubt these admit of more complete analysis and subdivision, and others may be added. We must bear in mind their entirely subjective character. They are strong in proportion to the size of the part of the brain with which they are connected, and not at all in proportion to the size or worth of the object of their desires. A little girl loves her headless or eyeless doll quite as much as she loves her children in after life. She will not change it for a new, more perfect, or better dressed one, any more than in after life she would change a damaged or spoilt child for a better one; her love supplies its imaginary attributes, and is stronger generally as it wants her love and assistance. It is precisely the same in her "first love;" the feeling is within herself, and transfers its perfections to the object, however wooden he may be intellectually,

* "Man's Nature and Development," p. 188.

† On "Force and its Mental Correlates," p. 43.

or one-eyed morally. It is the same with other feelings—a man is angry and oppositive, and looks upon all the world as at war with him. A peaceable citizen (as the story goes) venturing to address a Scotsman in this state of mind with the mild remark, "A fine day, Sir," was met with the rather unexpected reply, "And wha said anything against the day, I should like to know; you'd quarrel with a stone wa'." When feelings are much in excess, they constitute monomania—such as the homicidal mania, and a tendency to steal (kleptomania), which cannot be resisted, and all our feelings and emotions are occasionally in excess, and proportionally uncontrollable—in fact we are all a little cracked, though individually we are unconscious of it, and can only see the crack in others.

We must always bear in mind that these Propensities and Sentiments are blind impulses, requiring the Intellect for their guidance. They impel us to action by the pleasure we find in their gratification, or the pain in their denial; but they in no case tell us what to do. A mother's love for her child does not tell her how to treat it, nor does a love of right and justice tell us what is right and just. There is a strange jumbling together, in the minds of Physiologists and Psychologists, of the action of these faculties. The intuitive moralist says, that because a man has an instinctive love of justice, that love tells him what is just. Another says that there is no action of the feelings without an action of the intellect at the same time; whereas it is notorious that an irritable, combative person knocks a man down, and thinks and reasons about it afterwards; there is a blow and a word, not a word and a blow. And so of all the feelings—we often act upon their mere instinct or impluse; self-consciousness, or reflection, comes afterwards. Of course it is quite true that there is no self-consciousness, which means reflection on consciousness, without ideas or the action of the intellect: and probably that is all is that meant.

## THE INTELLECTUAL FACULTIES.

The Senses, although they have each their own peculiar sensation attached to them, of light, sound, smell, touch, and taste, yet must be regarded principally as the medium by which the brain is brought into action and ideas created. Take, for instance, our ideas of colour. Professor Tyndall tells us that "the atoms of luminous bodies vibrating, communicate their vibration to ether in which they swing, being propagated through it in waves; these waves enter the pupil, cross the ball, and impinge upon the retina at the back of the eye. The act, remember, is as real, and as truly mechanical, as the stroke of sea waves upon the shore. The action of the ether is communicated to the retina, transmitted thence along the optic nerve to the brain, and then announces itself to consciousness, as light." * * * "It would take 57,000 waves of violet colour to fill an inch, and as light travels at the rate of 192,000 miles a second, it would take 699 million of million of such waves *to enter the eye in a single second* to produce the impression of violet in the brain."* Thus it is the action of a portion of the brain, stimulated by the optic nerve, that gives us the idea of violet, or the nervous sensibility we call colour; and if the organ of the brain is not there we have no idea of colour, and many people with deficiency of this organ cannot distinguish some colours from others, or are even colour blind. What would the world be if every one had this deficiency of brain and all colour were taken away! Colour appears to be attached to certain forms, sizes, weights, in certain relative positions, order, and number, and the sense of these things—that is our ideas of them—are all equally dependent on the action of the brain, and if the brain is not there the ideas are not, and the ideas are vivid in proportion to the amount

* "Heat a Mode of Motion," pp. 235, 243.

of brain in each separate organ. Now it is absurd to suppose that there can be any possible resemblance between the wave motion, producing the 699 million of million knocks on the eye, and the nervous impression or idea we call violet, or the more or less millions of other knocks required to produce other colours. It is precisely the same with respect to all our other ideas: they are produced by the brain being set in motion by some mode of action from without, or one part of the brain sets another part of it in motion, inducing a set of ideas or a succession of consciousness, which we believe to be a world without; but as we have said, there can be no possible resemblance or analogy between such consciousness and the mode of action which produces it. The idea of a thing and the thing itself, or the idea and its cause, whatever that may be, are certainly not one and the same, nor can we conceive even how there can be any possible resemblance; for if even, as it may afterwards appear, the motion, or rather the cause of motion, and the idea it causes, may be a correlation of the same force, yet they differ essentially in form or condition, if not in essence. The world is created within us by the action of our own faculties. All we know is our own consciousness, and the modes of action that produce it.

These modes of action we classify, and show their relations to each other, and we call this Science.

Our "forms of thought" connected with, and dependent upon, certain portions of the brain, first, out of a simple and probably a single force, create what are called the properties and qualities of bodies, corresponding in our speech to the adjective—the form, size, weight, colour, order, number, and locality; another faculty gives unity and individuality to these properties, and creates the substantive, or individual—it gives the idea of the thing or person, while another faculty takes cognizance of its mode of action, corresponding to the verb; another traces analogies, and resemblances, and differences; another congruities or incongruities, upon which wit and humour depend; while another faculty traces out permanent relations of cause and effect, as opposed to mere sequence, and gives the sense of power upon which this permanence, or law as it is called, is supposed to depend. We have ideas also of space and time, and must, therefore, have faculties that create them; but however difficult it may be to conceive, they are not entities, but pure creations of the mind, and have no existence out of ourselves. "An entity," says Luke Burke, "must exist in some condition, it must also exist in some *where*, in some place, in some space. Place and space, therefore, are not entity, nor conditions or properties of entity, nor any reality whatever; they merely express the emptiness which entity fills or might fill. They are purely terms of negation." * * "Time, like space, is a term of relation merely, and has no reality corresponding with it."*

As substances may be so combined as to produce a sense of harmony and beauty, so also sounds may be so related and combined as to produce a sense of harmony of sound. Of the faculties to which these are related, the one has been called Ideality, the other Tune. We have also another faculty by which we are able to associate our feelings, emotions, and ideas with audible sounds, and to express them in what we call Language. Other creatures have also their languages, unfortunately unknown tongues to man. I say unfortunately, because man makes his ignorance the measure of their capacity, and what he loses by such ignorance is incalculable. By this simple action of all these faculties the world is created within us; and as an instinctive belief, as we have seen, attends the action of each faculty, we believe in its objective existence, and not in its mere ideal existence. That is, we believe that there is something in

* Principles of Ethnology.

bodies themselves corresponding to what we call colour; whereas there is merely a vibratory motion which communicates itself to the ether, which communicates itself to our brain; colour is therefore more the property of the brain than of the distant luminous object by which it is affected. And so with all our other ideas; as brains differ in relative size in different individuals, so do such persons differ in the clearness of their ideas, and their talents and dispositions; and as persons may be colour blind from want of brain, so may they be equally blind in Tune, in Reasoning Power, and in fact in all their other faculties. This is why people take such different views of things *—because their faculties, differing according to the size of the parts of the brain with which they are connected, they are differently affected by the forces without. But the mode in which each person is affected constitutes the truth to him.

Perception, Conception, Memory, Imagination, Judgment, are not themselves faculties of the mind, but mere modes of action of the primitive faculties.

The brutes possess these faculties as well as man, and some of them in a much higher degree than man. The instances of Reasoning we notice in some of them are probably mere Memory of what they have seen, or Judgment, for each faculty has an intuitive Judgment belonging to it, independent of Reason. If any of the brutes have reason, it is probably in a very minor degree, as some power of progres-

* Of course there are other reasons why people differ so widely in opinion besides their peculiar or subjective mode of viewing things. Objectively every question has a hundred sides at least in which it may be viewed. The ordinary observer perhaps sees only one or two; the man who studies the question may see fifty, and differs proportionally with the one who sees only two, and whose dogmatism and bigotry are generally in proportion to his ignorance; but public opinion is with the latter, he being in the majority—on the principle that nothing multiplied a sufficient number of times comes to something, or that a hundred fools make one wise man.

sion always attends reason or reflection. It gives the power of looking before and after—of anticipating events, and would not suit the brutes, for as Pope says:—

> The lamb thy riot dooms to bleed *to-day*,
> Had he thy reason, would he skip and play?

The intellectual faculties give us no *real* knowledge of anything—we know only the consciousness that external forces create in us through them; and even of the *real nature* of that consciousness we know nothing. Neither is it necessary that we should know more; all we require is as much knowledge, and of that kind, as will guide us towards the objects of our being, as indicated by the Propensities and Sentiments, the ultimate objects of which are our enjoyment, and the production of a generation to follow us when we are gone. Our knowledge is thus Relative, not Absolute.*

* "That we know matter only as relative to our own susceptibility of being affected by it, does not lessen the value of the knowledge of it which we are able to acquire; and indeed it is only as it is capable of affecting us that the knowledge of it can be of any direct and immediate utility. It would indeed be the very absurdity of contradiction to suppose ourselves acquainted with qualities which cannot affect us."—(Dr. Thomas Brown, "Philosophy of the Human Mind," Lecture 9.)

"How short soever our knowledge may be of a universal or perfect comprehension of things, it yet secures our great concernments, that we have light enough to lead us to the knowledge of our Maker and the sight of our own duties. * * As to myself, I think God has given me assurance enough as to the existence of things without me; since by their different application I can produce in myself both pleasure and pain, which is one great concernment of our present state. * * Our faculties serve our purpose well enough, if they will but give us certain notice of those things that are convenient to us."—(Locke.)

"What if I am indeed no other than this fine bodily instrument made sensitive to a thousand impulses—what if I am indeed this 'living lyre,' swept over by every wind, and tremulous to every ray of light—living lyre conscious of its own melody? I am still nothing

In the vast chain of Being between man and the world of animalculæ existing in a drop of water, no doubt there is an equally vast variety of Intellectual faculty, creating for each creature a world of its own, directing each to the objects of its being, and fitting each for a life of pleasurable enjoyment in its own peculiar sphere, however limited that sphere may be: making up probably in quantity of pleasurable sensations what it wants individually in quality.

less than that wondrous instrument that has converted *motion* into melody—the *thing* into a *thought*. Or, to change my metaphor, I am that sensitive mirror in which the reflected world becomes a conscious world, and knows itself as the creation of God."—(Thorndale on "The Conflict of Opinion," p. 359.

---

According to a recent estimate the population of the globe is about 1,228,000,000 souls. Of this number 552,000,000 belong to the Mongolian race; 360,000,000 to the Caucasian; 190,000,000 to the Ethiopian; 176,000,000 to the Malay, and 1,000,000 to the Indo-American. The annual mortality is over 33,000,000, or more than 60 every minute.

According to the calculations of Prof. Wappäus, the average duration of human life is 28.19 years in Austria; 30.60 years in Sardinia; 31.10 years in Prussia; 31.16 years in Saxony; 32.16 years in Bavaria; 34.72 years in Holland; 36.92 years in England; 38.35 years in Belgium; 40.36 years in France; 40.49 years in Denmark; 40.66 years in Sweden; and 43.64 years in Norway.

The number of languages spoken in the world amounts to 3,064. The inhabitants of the globe profess more than 1,000 different religions. The number of men is about equal to the number of women. The average of human life is about 33 years. One quarter die previous to the age of seven years; one half before reaching seventeen; and those who pass this age enjoy a felicity refused to one-half the human species. To every 1,000 persons only one reaches 100 years of life; to every 100, only six reach the age of 65; and not more than one in 500 lives to 80 years of age. The married are longer lived than the single, and above all, those who observe a sober and industrious conduct. Tall men live longer than short ones. Women have more chances of life in their favour previous to fifty years of age than men, but fewer afterwards. The number of marriages is in proportion of 75 to every 100 individuals. Marriages are most frequent after the equinoxes; that is, during the months of June and December, and depend not on love, but on the price of bread. Those born in the spring are generally more robust than others; so are those born from parents in middle life. Births and deaths are more frequent by night than by day. The number of men capable of bearing arms is calculated at one-fourth of the population.

"It is well known to all Frenchmen that in England our climate is so dreary and detestable that there are more suicides among us than in any other country of Europe. Statistics show, however, that for a long period of years the average number of persons who have destroyed themselves has been at the rate of 110 per million inhabitants in *La joyeuse France*, and only 69 per million in spleen-smitten Albion. During the past year no less than 5,011 persons committed suicide in France—4,008 men, and 1,003 women. Of these, 960 men and 407 women drowned themselves; 1,072 men and 835 women chose death by hanging; 498 men shot themselves, while but 5 women owed their deaths to firearms; 192 men and 113 women made use of the fumes of charcoal; knives were used by 176 men and 33 women; poison by 74 men and 44 women; 99 men and 55 women met death by jumping from windows, towers, &c.; 31 persons flung themselves in the way of trains. One individual starved himself to death. It thus appears that hanging is the most prevalent method of self-destruction in France, then comes drowning, then suffocation by charcoal, and lastly by poison. It is a well-known fact that, so far from decreasing, the number of suicides increases in a direct ratio to the spread of education and civilisation. In Prussia, the most highly educated country in Europe, the annual average of persons per million inhabitants who destroy themselves is 240—more than double the average of France, and nearly fourfold that of England. On the other hand, in Spain the average of suicides is only 14 per million inhabitants."—(*Pall Mall Gazette.*)

# CHAPTER III.

## MORALITY.

MORALITY is the Science of man's Duties. It concerns, in the first place, *Why* he should act in any particular way, or the Principles of Morality; and secondly, *How* he should act under the guidance of such principles, or the Practice of Morality.

### THE PRINCIPLES OF MORALITY.

In considering the question *why* a man should act in one way rather than in another, we have first to ask, is there anything *in actions themselves* that makes one line of conduct more desirable than another? I reply, certainly not. The differences that man has created in this respect are purely subjective, having no existence whatever without himself: in fact, the Moral World is as much the creation of the mind as we have shown the physical world to be. All actions are equally Necessary, that is, they are the result of antecedent irresistible force or cause, and therefore there can be nothing in the actions themselves to distinguish them from each other, or to make one action more a duty than another. Of course this is hard of belief; it has been, and still is, much disputed; we must give some little space therefore to its elucidation. The modern expression of this doctrine of "Philosophical Necessity," as it has been rather erroneously called, is to be found in the comparatively recent discovery of the Persistence or Conservation of Force. In deference to theologians and others bound to support special views, this doctrine of the Persistence and Correlation of Force has been hitherto confined to the physical forces, but Herbert Spencer and others have shown that it equally applies to Mind or mental force, and "that each manifestation of force can be interpreted only as the effect of some antecedent force; no matter whether it be an inorganic action, an animal movement, a thought, or feeling. Either this must be conceded, or else it must be asserted that our successive states of consciousness are self-created. Either mental energies, as well as bodily ones, are quantitatively correlated to certain energies expended in their production, and to certain other energies which they initiate; or else nothing must become something and something must become nothing." *

"The grand point permanent throughout all these considerations is that *nothing is created.* We can make no movement which is not accounted for by the contemporaneous extinction of some other movement." †

The more ancient expression of the same truth is given in various ways by various eminent authorities. I shall give a few only. Thus: "Nothing comes to pass without a cause. What is self-existent must be from Eternity, and must be unchangeable; but as to all things that *begin to be*, they are not self-existent, and therefore must have some foundation of their existence without themselves." ‡

"In no mind is there an absolute or free volition; but it is determined to choose this or that by a cause, which likewise has been fixed by another, and this again by a third, and so on for ever. * * * Human Liberty, of which all boast, consists solely in this, that man is conscious of his will, and unconscious of the causes by which it is determined." §

"Every action or phenomenon, so far as it produces an event, is itself an event or occurrence which pre-supposes another state wherein the cause is to be met with; and thus

* Herbert Spencer.

† "Heat a Mode of Motion," third edition, p. 499; Tyndall.

‡ Jonathan Edwards. § Spinoza.

everything that happens is but a continuation of the series, and no beginning *which occurs of itself* is possible: consequently, all the actions of the natural causes in the succession are themselves again effects." *

"Everything that exists depends upon the past, prepares the future, and is related to the whole." †

"Rejecting, then, the metaphysical dogma of free-will, and the theological dogma of predestined events, we are driven to the conclusion that the actions of men, being determined solely by their antecedents, must have a character of uniformity, that is to say, must, under precisely the same circumstances, always issue in precisely the same results." ‡

"The life of man is, therefore, like a stream of events or changes in linked sequence, flowing on as necessarily as the waters of Niagara. It is true that, in common language, the *will* is spoken of as the first cause of conscious thoughts and acts; but no acts of will (that is, of mental energising) can occur without *its* necessary co-existents and antecedents—that is, its causes; and such as these are, so will the act be. There is, in fact, no more a spontaneous act of will than there is spontaneous generation." § ‖

* Kant. † Oersted.
‡ H. T. Buckle. § Dr. Laycock.

‖ Suppose even that in a million people no two are alike, still, if that is the extent of their difference, all beyond that number would be like some in the million—one at least; so that in every 10 millions there would be 10 persons who would do exactly the same in similar circumstances. The fact is that none are exactly alike; still they are sufficiently so as to predicate actions with certainty in a given number. Experience or Statistics can alone point out what that number is. In a given number the crimes that will be committed, and the character of those crimes, are exactly known. The number of people who go mad, the number who commit suicide and the exact method, the number even who will put their letters into the post without any address, are certainly known. There is no chance, or contingency, or free-will anywhere: the same causes, under the

If then all actions are "Necessary," that is, if each manifestation of force, mental and bodily, can be interpreted only as the effect of some antecedent force, (which the persistence

same conditions, always produce the same results. In the Registrar-General's Annual Report there are certainly some things that look like chance. For instance, one person died through terror produced by a thunder-storm, one from the sting of a wasp, one from the bite of a donkey, and one was choked by swallowing a set of artificial teeth. A good many people may die from the bite of a donkey—witness the lamented death of the poor poet Keats; but the last-mentioned cause of death perhaps is not sufficiently general to assume the form of a law at present, and it may not be known with absolute certainty how many people will die next year from swallowing their teeth—at present they are too expensive to make the liability very great. The Registrar of Killin mentions that an old bachelor who died at the age of 91 years "five or six years ago cut six new teeth, which he said were quite serviceable and as sharp as lancets."—(The Registrar-General for Scotland, 14th Annual Report.) It certainly would be a great convenience, and save considerable expense and, as it appears, danger, if nature would provide a set of teeth for our second childhood, as it does for our first; but as this gentleman was 91 years old and a bachelor, I fear we have not much to hope for from "Natural Selection" in this case. It may sound like a paradox, but it is nevertheless true, that all History is false, but all Fiction is true. It is almost, if not quite impossible, to say truly what happened to *particular* individuals at some *definite specified* time and place, but it is probably equally impossible to state anything in fiction that has not occurred, and that is not most likely occurring at that very moment, somewhere in the world. Sixty people die every minute and in the 12 hundred millions of the human race—as like causes produce like effects—there is probably nothing we can imagine that is not taking place many times over somewhere during that minute. The motives that move us are similar, constant, and in constant action. Young people fall in love, and mothers are certain there are not, and never were, such darlings in the world as their children &c., and they think it all new and quite original; but the same thing is happening all the world over, and has been from time immemorial. There are the same murders, the same suicides, the same number of fools putting their letters into post without any address; Babbage could do it all by a machine.

of force implies,) and "free-will" is consequently a delusion, whence does this delusion arise? The freedom universally recognized is the power which every creature has, in common with everything else, organic and *inorganic*, to act in accordance with its nature or the laws of its being, when free from external constraint; which power, however, equally in mind and matter, is derived from without. Our consciousness it is that is guilty of the delusion in this case, as in so many others; the fact being that the persistence of force does not appear in consciousness, but only its correlation. The discussion on this much-vexed question would not have continued so long, and be yet undecided, if there were not some misunderstanding of the terms in which it is expressed. Persistent force or cause is always from without, but its mode of action or manifestation depends upon the nature of the substance or organ through which it passes, and when this action is unconstrained—that is, not interfered with or interrupted—it is called free. But everything has a nature of its own, and its mode of action is determined by that nature. In mental action we are thus *compelled* by our nature to do things *voluntarily*. This has been called a *free necessity*. Thus it is said, "God is free because He acts from the necessity of His own nature."* But everything has this freedom, which is no freedom at all, as everything, and everybody, and every mind, *necessarily* acts definitely in accordance with the laws of its nature or being. It is this which has been mistaken for freedom, as distinguished from the condition in which some impediment has been imposed to such action. Thus a thing is said to be free when it is determined to action by itself alone; but that self, whatever it may be, acts *necessarily* in accordance with the laws of its own nature. It is thus that people delude themselves into believing both in free-will and necessity, or in *free necessity*. It is necessity, not freedom, that increases as we ascend in the scale of being.

* Spinoza.

The flight of a swallow seems freest of all, and yet every devious motion has a cause, and might be calculated by superior intelligence. There are a hundred ways open to man by which he may go *wrong;* but according to the ordinary conception of God, there is but one way open to Him: He must always necessarily do that which is *right*, and only that. "Outward things cannot compel the gods," says Seneca, "but their own eternal will is a law to themselves." Seneca also says—and the remark illustrates what is commonly meant by freedom—"God is not hereby less free, or less powerful, for He Himself is His own necessity." A man feels that he is at liberty to do as he pleases when he is free from external compulsion; but what he pleases always depends upon the nature of his bodily and mental constitution, or, as Buckle puts it, "all the vicissitudes of the human race—their progress or their decay, their happiness or their misery, must be the fruit of a double action; an action of external phenomena on the mind, and another action of mind upon the phenomena." If we have not Freedom, then, what have we? Law—law in Mind as in Matter. Man always acts in accordance with the laws of his being, and if we would control him we must ascertain what those laws are. He is included in the realm of Science. Mr. J. S. Mill says: "The conviction that phenomena have invariable laws, and follow with regularity certain antecedent phenomena, was only acquired gradually, and extended itself as knowledge advanced, from one order of phenomena to another, beginning with those whose laws are most accessible to observation. This progress has not yet attained its ultimate point; there being still one class of phenomena (human volitions) the subjection of which to invariable laws is not yet universally recognised. * * At length we are fully warranted in considering that law, *as applied to all phenomena* within the range of human observation, stands on an equal footing in respect to evidence with the axioms of geometry itself." Accordingly, we find our most advanced physiologists and psychologists now recognising this truth. Thus Dr. Henry Maudsley says: "If it were possible for any one to enter thoroughly into the inmost character of another person, and to become exactly acquainted with the moving springs of his conduct in his particular relations of life, it would be possible not only to predict his line of action on every occasion, but even to work him, free-will notwithstanding, like an automaton, by playing on his predominant passion, interest, or principle. * * It would be possible to foretell with certainty his mode of thought and conduct under any given circumstances. * * Necessity is, in truth, confessed in every deliberation and in every act of our life." *

It cannot be, then, in actions themselves that we are to look for that which constitutes them either good or bad, as they are all *equally* necessary—all the effect of an efficient cause; and under the circumstances they *must* happen. Thus of the various kinds of killing and the various kinds of love-making which mankind approve or disapprove: the acts themselves may be precisely the same, and yet one is

* "The Physiology and Pathology of Mind," pp. 183, 143, 141.

This question of "freedom of will" is closely analogous to the question as to whether the sun went round the earth or the earth round the sun in the 15th century. All men then believed the sun to go round the earth, for did they not see it do so every day! And do not all men now feel that they are free! A young gentleman in Galileo's time, going up for examination, was asked whether the sun went round the earth or the earth round the sun, and, in deference to the public opinion of his day, said, "Sometimes one and sometimes the other." And such is the present attitude of men of science to this question of Necessity or Freedom of Will. President Edwards, the clearest and most powerful reasoner of his or any other day, in accordance with the behests of his theological creed, held both doctrines to be true, and our men of science, in obedience to an equally imperative public opinion, also profess to hold both doctrines, or that volitions are sometimes *necessary* and sometimes not—"sometimes one and sometimes the other."

called murder, another merely homicide; the one marriage, another adultery. If, then, there is nothing in actions themselves, they being all mere modes of motion, governed by law, what is there to distinguish one action from another, to make it good or bad, our duty or not? This question can only be answered by another, for how can we test actions except by their reference to some end? The question, therefore, is, what is the end or object of all action? and an action may then be described as good or bad, our duty or not, as it best tends to promote that object. It seems to me that there is but one answer to this question, namely, that the sole end and aim of existence and of all action is pleasurable consciousness or happiness. Reasoning *a priori*, we can come to no other conclusion. Thus a world without consciousness, of which no being was percipient, would be the same as no world at all. Now consciousness is a feeling, and a feeling must be either pleasurable or painful, however slightly so. No feeling would be no consciousness; a painful feeling would be an undesirable consciousness; and consequently, there is nothing left worth having but the pleasurable consciousness. Life! what would be the good of that if it were not enjoyable? Knowledge! what is that but a record of how things affect us, pleasurably or painfully, —a mere register of the few modes of action that are capable of setting our brains in motion through the senses, and which is different in each creature, as that brain is modified to suit the various requirements of its being? In fact it is impossible to conceive of any other object in creation but Happiness, either present or to come. Practically and intuitively all actions are tested by their conduciveness to this end. What is the use of it? what is the good of it? mean ultimately, Will it make anybody happier? Actions that were "free" would not only have no virtue in them, because of that freedom, but the gift would be a fatal one, "a mockery, a delusion, and a snare," if that freedom could be used by the majority of mankind to damn themselves to all eternity, as some very good people suppose. The predominance of what are called the higher feelings over lower ones—of the unselfish over the selfish—is only desirable because it makes more people happy. Heavenly aspirations, when they take the direction of asceticism, are only sacrificing happiness now that more may be had hereafter; it is a mere exchange of worldliness for other worldliness. A future state of rewards and punishments is only carrying utilitarianism into another world.

How is it, then, that on a question apparently so self-evident there should be still two schools—the Intuitionists and the Utilitarians? It is from ignorance of human nature; from the too prevalent want of a true psychological system. None of our faculties seek happiness for its own sake, although happiness results from their attaining their ends. A woman loves and tends her children, and it makes her happy to do so; but she does not love and tend them that she may be happy. And so of all our other motives to action: they none of them have happiness directly for their object, but if pleasure did not follow from the pursuit, or pain from the neglect, they would cease to be motives. It is in this sense only that we can accept Jeremy Bentham's dogmatic assertion that "No man ever had, can, or could have a motive differing from the pursuit of pleasure or the avoidance of pain." No doubt these are the ultimate springs of all action, although they are hidden from us. Again he says, "the first law of nature is to seek our own happiness." Now we never seek our own happiness, but rather the objects of our desires, and we find happiness in doing so. A man seeks to do what is right, and has pleasure in doing it: he seeks to make others happy, and if he succeeds that makes him happy; but his own happiness, in either case, is the last thing he thinks of. But the error of the Intuitionists is here: because there is an innate *love* of right in the abstract, and a man has pleasure, therefore, in doing what is right, consequently they infer he knows what *is* right. On the other hand, the Utilitarians, seeing that men, in all time, have everywhere differed in opinion as to what is right, have come to the equally erroneous conclusion that there is no instinct or faculty that makes us desire or love right for its own sake. It is true we have certain likes and antipathies, amounting in some cases to idiosyncracies; it is true also that certain things naturally, and at once, approve themselves to us as right or wrong, but that is no reason why they should be right or wrong; and as all men differ with respect to them, or at some time or other have differed, we do not the less require some external standard by which to test our feeling on the subject; and if there is a fact in history well established it is the gradual formation of the Conscience on the external standard of utility, that utility not being the mere expression of the intellect, but of all our intuitions of feeling as well as reason. So the love of our fellow-creatures does not teach us how best to promote their interests. It is true, however, that in proportion to our love of right and of our fellows, do we earnestly seek the right and that which will do them good,—still all our instincts are mere blind impulses, requiring the guiding hand of reason to lead them to their objects. The love a mother has for her child does not teach her how to treat it; and the love of truth and right, and the preference we naturally give to some actions, does not tell us what is true and right, for through all time "what one nation views as a crime another praises as a virtue; what one nation glories in as a legitimate pleasure, another reprobates as a shameful vice: there is scarcely a crime or vice that has not been exalted in a religious observance by one nation or other, at one period or other of the world's history." * The fact is, the Conscience, like all the other intuitions or instincts, is transmitted experience of utility—it is "the gathered-up

* "The Physiology and Psychology of the Mind," p. 157. Dr. Maudsley.

experience of by-gone generations, transmitted to us by inheritance," requiring to be brought by the intellect up to the light of the present age. The kitten, immediately it opens its eyes, sets up its back at its natural enemy, the dog; the cow tosses its head at every dog it sees, because bull-dogs in previous ages "baited" its sire; and a man's conscience would toss everything that previous ages has proved to be injurious to his race and therefore wrong: it is a power that enables him to recognise intuitively actions whose tendency in times past was to produce happiness or misery.

But suppose we admit, contrary to all that experience teaches, that we have some internal monitor that on all occasions tells us what is right, why should we *do* right, when we do know? We are all aware how easy it is to know what is right and not to do it. The answer is, that if we did what was right it would give us pleasure; or if we did not, conscience would give us an awful pinch; so that it is pleasure or pain, that is utility, that must set conscience in action, supposing even that it is a perfect monitor, instead of a mere blind impulse. Conscience, after all, is only one among other motives to action; there are religion, law, benevolence, fear of consequences here and hereafter; and the weighing of consequences is utilitarianism.

Mr. Darwin bases Conscience upon Sympathy, by which he really means sympathy with these other motives—religion, law, benevolence, and the others. But sympathy is no special faculty, but like memory it belongs to every faculty. Thus, a mother sympathises with maternal feeling in another in proportion as her own love of offspring is strong, and so of all our other feelings. A man with small conscientiousness has little sympathy with truth and justice; but still if he has other feelings strong, his conscience may include them, and he may not only sympathise with the love of children and friends, and of his fellow-creatures, but with self-respect, self-command, deference to others, a chivalrous sacrifice of self, &c. A love of right includes sympathy with all these higher motives, but conscientiousness is the love of truth and justice, and they properly only allow these other sympathies to be exercised in accordance with truth and justice. Conscience, with Mr. Darwin, evidently means the dictates and predominance of the moral feelings, although he does not distinctly see what those are. Still, this highest conscience is built on utility, for, as he tells us, "No tribe could hold together if murder, robbery, treachery, &c., were common; consequently such crimes, within the limits of the same tribe, 'are branded with everlasting ignominy,' but excite no such sentiments beyond these limits. * * It has even been recorded that an Indian Thug conscientiously regretted that he had not strangled and robbed as many travellers as did his father before him."* It is quite true that man's conscience, or sense of right and wrong and duties to others, *begins* in the family, and gradually extends itself to the tribe, then to the human family, then to the whole sensitive creation, and lastly to the Ruler of the Universe, as ultimately he learns to regard the Natural Law as God's will and mode of action.

But why, on Utilitarian principles, should a man do right when apparently it is against his individual interest on some special occasion? To take an extreme case, why should Regulus, for instance, return to his barrel of spikes? There are three ways in which we may see distinctly Regulus would have been a gainer. In the first place, if he had not possessed all the higher attributes of mind, he never would have returned to his spikes, and he had, therefore, been a great gainer by the possession of the highest class of mind all his life; for

Better fifty years of Europe than a cycle of Cathay.

In the second place he had been a gainer by the higher

* Vol. 1, pp. 93, 94.

moral atmosphere in which he had lived all his life, consequent upon others having previously acted upon the same high sense of honour; and, lastly, if he had not returned, the sense of having acted unworthily would have made life to him not worth having—a pain instead of a pleasure.

This shows that the calculation of right and duty should be based on the general happiness, and that in the long run will best secure the happiness of the individual, although it must be admitted Nature pays small regard to individuals, and never hesitates to sacrifice the interests of the individual to that of the race. It is the province of moral science to lay down general rules, based on the public good, which people should follow implicitly, without calculation of the specific consequences of their own actions to themselves individually. Happiness depends upon *habits of mind*, and the great object of moral training is to make the moral law automatic—that is, we should be trained to do at once unconsciously what is right, and good, and true, and generous, without thinking at all about it, much less stopping to calculate consequences. Undoubtedly that is the happiest man who if his friend fell into the water would go in after him at once, whether he could swim or not, and who, if he had given *his word* to return even to a death of torture, would certainly do it.

But some people, knowing that after all happiness is but the aggregate of pleasurable sensations, despise poor happiness so compounded, and think it beneath them to live for it; and they tell us the object of life is progress, or knowledge, or evolution, or moral perfection, or spiritual growth, aspiringness, and sublimation; but they do not tell us what use these things are if they do not lead to happiness. As one class despise mere pleasure and pursue happiness, others discard happiness and contend for blessedness; but this is all talk. No one has yet shown us anything better than happiness, and all pursue the line of conduct that they think most likely to lead to it.*

* Bentham says: "The various systems that have been formed concerning the standard of right and wrong, may all be reduced to the principle of sympathy and antipathy. One account may serve for all of them. They consist all of them in so many contrivances for avoiding the obligation of appealing to any external standard, and for prevailing upon the reader to accept of the author's sentiment or opinion as a reason, and that a sufficient one, for itself. The phrases are different, but the principle the same."

See also Bentham's note on this passage: "If, therefore, a man wants to know what is right and what is wrong, he has nothing to do but come to me," because, he says, he has a moral sense or conscience, or a common sense, or an understanding, or a rule of right, or an intuitive knowledge of the Fitness of Things, or of the Law of Nature, or of God's Will, which always tells him what is right, but which in truth is nothing more than his *ipse dixit*.

Lecky is a professed Intuitionist, yet the whole of his very valuable work, "History of European Morals," points to the opposite conclusion—viz., that the Conscience or Sense of Right varies, and has varied at all times and in all countries; proving the necessity, beyond all doubt, for an appeal to some external standard. The Conscience has been a thing of very slow growth, each age applying a different external standard in accordance with its advance in civilisation and refinement. Much requires to be done in the present day. Half the philanthropy in the world is eleemosynary, and has been proved to be doing mischief. But let us follow Mr. Lecky, to see how far Conscience may be trusted as a standard of right. He tells us (vol. 2, p. 19) that Nature does not tell men that it is wrong to slay without provocation his fellow-man. Look at the Hebrew law: "And if a man smite his servant, or his maid, with a rod, and he die under his hand; he shall be surely punished. Notwithstanding, if he continue a day or two he shall not be punished: for he is his money." Exodus, chap. xxii., v. 20-21. Here we have murder and slavery condoned, or rather "the *right* of every one to whop his own nigger" to death, if he does not kill him at once, recognised by Law, by God's chosen people.

Aristotle, and we have none greater among the ancients, taught that Greeks had no more duties to barbarians than to wild beasts (vol. 1, p. 241). Aristotle, however, only reflected the Consciences of his age. As Lecky says, "At first, the range of duty in the family, the tribe, the state, the confederation. Within these limits every man feels

*A priori*, then, there would seem to be nothing better than happiness or pleasurable consciousness, and whatever may be said in theory, it is evident that in practice all the

himself under obligations to those about him; but he regards the outer world as we regard wild animals, as beings upon whom he may justifiably prey. Hence Brigand and Corsair, in the early ages of society, conveyed no notion of moral guilt" (vol. 2, p. 271). The public conscience had not greatly advanced even up to our Elizabeth's times if we may judge from the adventures of Drake and Sir Walter Raleigh. Also during ages of the Christian era, according to the popular belief, all who differed from the teaching of the orthodox lived under the hatred of the Almighty, and were destined after death to an eternity of anguish, and no moral or intellectual excellence could atone for their crime in propagating error (vol. 1, p. 206). Persecution to the death was therefore both logical and right.

Aristotle also not only countenanced the practice of abortion, but even desired that it should be enforced by law, where population had exceeded certain assigned limits (vol. 2, p. 22). In the present day we think it *right* to allow half the children that are born to die before their fifth year, for want of proper sanitary and other arrangements. There may, however, be some nice distinction between Aristotle's method of killing and this allowing to die.

The *right* of conquerors to massacre their prisoners was universally admitted (vol. 1, p. 302).

The Gladiatorial Shows of the Romans were originally religious ceremonies at a funeral; then we have Cæsar and Pompey competing with each other by greatly multiplying condemned criminals, who killed each other at these shows, till, as Mr. Lecky says, the main amusement of all classes was this spectacle of bloodshed, of the death, and sometimes of the torture of men, "and all this continued for centuries without a protest" (vol. 1, p. 287). Nero, on account of his munificence in this respect, was probably the sovereign who was most beloved by the Roman multitude (p. 298). Cicero, Seneca, Plutarch, &c., "of all the great historians who recorded them (the shows), not one seems to have been conscious that he was recording a barbarity" (p. 308); and Suetonius quite thought he was called upon to apologise to the people of Verona for refusing them "a show." "After so general a request," he said, "to have refused would not have been firmness—it would have been *cruelty*."

To come a little nearer to our own time, Mr. Lecky tells us that Sir Thomas Moore was famous for his skill at Cock-throwing, which

world admits this, and that that is right which best tends to promote it.

Reasoning *a posteriori*, we find that as far as we are able to trace and understand all that has been going on, from the beginning of the world, there has been a gradual preparation for the largest possible amount of enjoyment, until the whole earth has been covered with one net-work of nerves throbbing with pleasurable sensation. If man does not see this it is

was a famous English game, and is said to be connected with the denial of St. Peter. As Sir Charles Sedley said :—

Mayst thou be punished for St. Peter's crime,
And on Shrove Tuesday perish in thy prime."—(Vol. 2, p. 174.)

The national conscience does not appear to have spoken very loudly either as to the cruelty or the still *greater* injustice.

In Queen Elizabeth's time a decided preference was given to Bear-baiting over Shakespeare's Plays. Dr. Parr was fond of Bull-baiting; Windham and Canning strongly defended it, and the late Sir Robert Peel argued strongly against its prohibition.

But if Conscience, or the Sense of Right, seems to have been very dead in some directions, it appears to have been very lively in others. Thus, the "effusion of blood" was unscriptural, therefore the Inquisition burnt to death all those who differed from them on theological points; in obedience to the same command, we are ordered to abstain from the use of the sword, so certain sects beat to death those who differed from them with massive clubs. Unchaste vestal virgins were buried alive, because they were so sacred that it was unlawful to lay violent hands upon them (vol 2, p. 44); and a woman was actually executed for undressing before a statue of Domitian (vol. 1, p. 275).

The same delicacy of Conscience in Religious matters has come down to our own day. Thus the King of Burmah thinks, as before mentioned, it a damnable offence to kill anything, but he does not object to *inferior* people killing animals and endangering their souls that he may have meat to eat.

Dr. C. J. Vaughan seems to think that in the burial service, which is read alike over saint and sinner, the words "sure and certain hope," &c., should be read *ironically* over a notoriously wicked soul! (see *Athenæum*, March 21, 1868, p. 414); and perhaps, also, we ought to wink! No doubt this will quiet many uneasy consciences.

from the narrow, perverted, purely subjective way in which he regards things. He has a pre-established theory to support. He sets himself up as the centre of all creation, at least in this world, and regards everything from the narrow point of view of his own individual interest or that of his race, whereas it is the nervous system of the world that nature cares for, and not the individual creatures of which that system is composed.* As we have seen, all action results from pre-existing force; and as all action, therefore, is on a par—being all equally necessary—all creatures must be so too, and we have one great throb of the heart of nature surging through the whole, and pleasure and pain regulating the conscious world as attraction and repulsion do the physical. As in the individual body there is one unceasing action of renovation and decay, an intense stream into which new matter is for ever flowing, and from which the old is ever moving out, and the intensity of life of the whole body is in proportion to the death of the parts; so in the great body of sensitive existence the intensity of life and enjoyment is in proportion to the degree in which new life is

* Fleas there are that bite men,
And fleas that bite those fleas again;
These little fleas have fleas that bite 'em,
And fleas bite fleas ad infinitem.

And no doubt the biter has more pleasure than the bitten has pain. Man is not *necessarily* included in this scale. If he keeps himself clean, and has the additional health and happiness that cleanliness will give to him, he is free from such inroads; but if he does not he becomes the feeding ground for other creatures, and nature thus restores the balance of enjoyment. Probably this applies to the internal parasites that are so great a difficulty with those who can see no end in life, no purpose or design in creation. These creatures point to some obstruction in the bodily constitution—some impediment to more perfect enjoyment that requires to be removed. Animals live upon one another, thus avoiding "the slow and snake-like life of dull decay," and the pang that deprives them of life is a mere *flea bite* to the long period of enjoyment they have previously had.

introduced, and the old, and comparatively unconscious, carried out. Life and Birth, Death and Decay, are the means by which the whole sensitive world is kept ever going and vigorous and capable of the largest amount of enjoyment. Of course this does not take place without a great outcry on the part of those who, having had their day, and having little capability of enjoyment left, would yet stop the way of the young and vigorous, who are equally entitled to their day, and whose term of life and turn for enjoyment have now come.*

But there are pains in the world as well as pleasures,—misery as well as happiness. We call this pain "evil," and as God can produce only good, man makes a Devil as the author of the evil—although God is supposed to have made the Devil in the full fore-knowledge of the consequences of what he was doing. But *is* Pain an evil? If the pleasures decidedly predominate, and the pains are absolutely necessary

* Neither must we be misled by the word "individuality," because it appears from the many facts and arguments in this book that our personalities are not so independent as our self-consciousness leads us to believe. We may look upon each individual as something not wholly detached from its parent source—as a wave that has been lifted and shaped by moral conditions in an unknown, illimitable ocean. There is decidedly a solidarity as well as a separateness in all human, and probably in all lives whatever; and this consideration goes far, as I think, to establish an opinion that the constitution of the living universe is a pure Theism, and that its form of activity is what may be described as co-operative. It points to the conclusion that all life is single in its essence, but various, ever varying, and inter-active in its manifestations, and that men and all other living animals are active workers and sharers in a vastly more extended system of cosmic action than any of ourselves, much less of them, can possibly comprehend. It also suggests that they may contribute, more or less unconsciously, to the manifestation of a far higher life than our own, somewhat as—I do not propose to push the metaphor too far—the individual cells of one of the more complex animals contribute to the manifestation of its higher order of personality.—("Hereditary Genius," p. 276, by Francis Galton, F.R.S.)

to the production and protection of those pleasures, then pain is not an evil, but a good; although the sooner we are able to do without its aid, and the less we have of it, the better. Dr. Neil Arnott says: "Man, like other warm-blooded animals, must always have the four prime essentials to life—*fit air, warmth, aliment, and rest after action;* and that to obtain the pleasure from using, and to escape the pain from wanting these, are the chief motives to his voluntary action."* The "pleasure from using," however, would often not be enough to insure voluntary action without the pain, and life, from pure forgetfulness, would not be maintained. The pleasures of appetite are all very well, but if it were not for the pains of hunger we should often forget the necessary "aliment." Nature tries pleasures first, and then if we do not do what is necessary for our existence and even well-being, she gives us an awful pinch somewhere. Is this an evil or a good?

Pain is necessary, then, as a sufficient and efficient motive to action where pleasure alone would not insure even continued existence. It is necessary also as a guardian to that delicate system of organisation, which is "so fearfully and wonderfully made," upon which our life and happiness depend. No surveillance, no admonition, would so well keep a child out of the fire as the pain that comes from being burnt, and the world is full of things that have a tendency, if not an equal tendency, to injure us, and of whose injurious qualities we know nothing till pain warns us of them. Pain and pleasure guide our choice among the infinitely varied constitutions of the things around us; and we are *obliged* to attend to the pain, if not to the pleasure. If the nerve is destroyed in the hand and foot, which nerve tells us when we are injuring it, the hand and foot, no longer having the protection of pain, are often destroyed too, as is so generally the case where the

* "Survey of Human Progress."

nerve is cut in horses' feet. The more delicate the organisation, and consequently the higher the susceptibility to enjoyment, the greater is the liability to injury. The eye, the most important inlet to the mind, is the organ most easily damaged. "The nerve of the coat of the eye is sometimes injured, and is no longer sensitive to the dust which adheres to the ball. Then the lid is not excited to wink, or the tears to flow. The particles which are carried into the eye cease to pain, and being allowed to remain, they set up inflammation, and the inflammation renders opaque the transparent covering through which the light flows. Blindness is the result, and the sight itself is found to be dependent upon the refined sensibility of the outer membrane." *

Pain is thus the strongest motive we can have, the best guardian, and the wisest schoolmaster. It constitutes that Necessity which is the mother of invention, and is the base of all Progress. No doubt we suffer, and have always suffered, from being a little too thick upon the ground; but as Herbert Spencer says, "From the beginning, pressure of population has been the proximate cause of progress."†

Pain then, if it is an evil, is a necessary one, and we can no more do without it than a child can do without its nurse. No doubt the child objects to the necessary constraint, and considers its nurse an evil. Pain or punishment follows error whether we do wrong voluntarily or not. No distinctions are made with reference to moral guilt, or actions that are supposed to be free. Punishment follows on all alike, the object of all punishment being the good of the persons punished—to teach them that they have done wrong, and to oblige them to do better in future. "Zeno, the founder of the Stoics," says Toplady, "one day thrashed his servant for pilfering. The fellow, knowing his master was a fatalist, thought to bring himself off by alleging that he was destined

* "Sense of Pain in Man and Animals." *Quarterly Review*, Jan., 1858.

† "Principles of Biology," vol. 2, p. 506.

to steal, and therefore ought not to be beat for it. 'You are destined to steal, are you?' answered the philosopher 'then you are no less destined to be thrashed for it.' But the necessitarian is not a fatalist; what therefore is more to the point is, that a school boy, knowing his master's opinion that no action could possibly have been otherwise *under the circumstances*, pleaded, for some great fault of which he had been guilty, that he could not help it. 'I know you could not,' replied the master, 'I do not blame you, I only pity you; and when I have given you a good flogging, the remembrance of it, and that it must always recur if the fault is committed, will enable you to help it in future.'" To be *saved* from punishment that is for our good is to do us a mischief, and yet that is what all are constantly praying for. Punishment that is not for our good is revenge, and to suppose that God, either in this world or in any other, revenges Himself upon His creatures, is a gross libel upon our Creator. To suppose that God can have any other object than the good of His creatures, or any attributes that are opposed to it, such as the vindication of His sovereignty, or authority, or supreme justice, is also a libel.

To the Necessitarian there is no such thing as sin and evil; only pains and pleasures. The distinction between physical and moral evil cannot be maintained. Sin, vice, and moral guilt are only evils from their tendency to produce pain and misery. Repentance and Remorse are foolish regrets over what certainly cannot now, and under the circumstances could not then, possibly have been otherwise, and are to be indulged only so far as they furnish a sufficient motive for our doing better in the future. As we get wiser and better we shall be more and more able to dispense with our nurse and schoolmaster, Pain. Probably there is no world altogether without pain, as there is no finite creature not liable to error; and pain, as far as we know, is the best, and perhaps only, corrective of that error. In a world in which there was no pain, it is probable there would be but little pleasure; certainly in those that have been created by ourselves, to supply the supposed deficiencies of our Creator here, life would be very slow, and nothing would so certainly take us back to the savage state as the imitation on earth of what we suppose to be the order in heaven.* In an island in the Pacific a little heaven on earth was supposed to be formed by the descendants of the mutineers of the "Bounty," and Mr. Edward Maitland has given us a description of their present condition. He says: "The principal characteristic of the islanders was their languid indifference to everything. They had never been forced to take trouble about anything, and they knew of nothing worth taking trouble about. It had cost them little toil or ingenuity to provide for all their bodily wants, and of mental wants they had absolutely none. They had never been allowed to know that there were two opinions in the world, even about religious matters. And so, without object of desire or ambition, without necessity of making a choice between Better and Worse, Right and Wrong, with no lack in the present, and secure of the future, they existed in a state of listlessness, more like vegetables than beings endowed with high faculties for happiness, for intellectual dominion, for moral excellence. And all I learnt only confirmed me in my idea of the meaning of man's history; that life is a force which has to be disciplined and educated by experience: that without temptation there can be no virtue, without pain no pleasure, without evil no good, without freedom no progress."†

* A little girl being threatened with heaven, which she supposed was to be all Sundays and collects, remonstrated and said, "Oh, but Ma, if I am very good I may have a little devil to play with sometimes, mayn't I?"

† "The Passion for Intellectual Freedom," p. 30, a Lecture delivered before the Sunday Lecture Society, Jan., 1871; published by Thos. Scott, Ramsgate.

Natural Selection.—The views contained in this Chapter were published by me some thirty years ago. Since then the works given to the world by Mr. C. Darwin and Mr. A. R. Wallace on Natural Selection, or Survival of the Fittest, have illustrated them in practice, and in this form they have obtained a wide acceptance.

It has been clearly shown that among animals, even among the least prolific, the tendency is to increase greatly beyond the means of subsistence, and that consequently, notwithstanding this great increase, the animal population of the globe is about stationary. This stationary state is effected mainly by animals living upon each other, one fitting comfortably into another—the eater by this arrangement probably having more pleasure than the eatee has pain. The consequence of this scarcity of food and of this necessity for defence is a struggle for existence, in which all the powers, bodily and mental, are exercised to their greatest stretch. Come, look alive! says Nature; and it is a struggle for dear life itself. We presume there can be no stronger motive for the exertion of all the energies. A system on which all are eating up each other, and on which the weakest always go to the wall, does not look as if it were based upon either love or justice; but it admirably promotes the largest amount of enjoyment, by developing to their fullest extent all the faculties upon which enjoyment depends, and keeping the world full of those only who are the most capable of enjoyment.

In this struggle, and under this pressure, Useful Varieties tend to increase, and useless ones to diminish, and thus Superior varieties ultimately extirpate and take the place of the original Species. "A wild animal," Mr. Wallace remarks, "has to search, and often to labour, for every mouthful of food—to exercise sight, hearing, and smell in seeking it, and in avoiding dangers, in procuring shelter from the inclemency of the seasons, and in providing for the subsistence and safety of its offspring. There is no muscle of its body that is not called into daily and hourly activity; there is no sense or faculty that is not strengthened by continual exercise. The domestic animal, on the other hand, has food provided for it, is sheltered, and often confined, to guard it against the vicissitudes of the seasons, is carefully secured from the attacks of its natural enemies, and seldom ever rears its young without human assistance. Half of its senses and faculties become quite useless, and the other half are but occasionally called into feeble exercise, while even its muscular system is only irregularly brought into action."* No doubt, consequently, as we learn from Darwin, "the brains of domestic rabbits and hares decrease in bulk, but increase in length, under domestication."† The self-protecting organs, Combativeness, Destructiveness, Secretiveness, Cautiousness, which give breadth to the brain, are diminished, and the Social, which give length, are enlarged. Man has only to apply these principles to himself, for the laws in his case are equally operative. He would deteriorate if deprived of the lower motives to action, if those that peculiarly distinguish him as man were not strong and active. A prize pig is a very different animal to a pig shaking down the acorns to its little ones in its native forest; and our aldermen should take a lesson! Mr. A. R. Wallace thus summarises his facts and inferences:—

"*A Demonstration of the Origin of Species by Natural Selection.*

| *PROVED FACTS.* | *NECESSARY CONSEQUENCES (afterwards taken as Proved Facts).* |
|---|---|
| RAPID INCREASE OF ORGANISMS. TOTAL NUMBER OF INDIVIDUALS STATIONARY. | STRUGGLE FOR EXISTENCE, the deaths equalling the births on the average. |
| STRUGGLE FOR EXISTENCE. HEREDITY WITH VARIATION, or general likeness with individual differences of parents and offspring. | SURVIVAL OF THE FITTEST, or Natural Selection; meaning simply that on the whole those die who are least fitted to maintain their existence. |

* "On Natural Selection," p. 38. † "Descent of Man," vol 1, p. 146.

SURVIVAL OF THE FITTEST. CHANGE OF EXTERNAL CONDITIONS, universal and unceasing.—See 'Lyell's Principles of Geology.'

CHANGES OF ORGANIC FORMS, to keep them in harmony with the Changed Conditions; and as the changes of conditions are permanent changes, in the sense of not reverting back to identical previous conditions, the changes of organic forms must be in the same sense permanent, and thus originate SPECIES."*

Mr. Darwin also thus summarises: "It may be metaphorically said that Natural Selection is daily and hourly scrutinising, throughout the world, every variation, even the slightest; rejecting that which is bad, preserving and adding up all that is good, silently and insensibly working whenever and wherever opportunity offers, at the improvement of each organic being in relation to its organic and inorganic conditions of life." He also says, "The term general good may be defined as the means by which the greatest possible number of individuals can be reared in full vigour and health, with all their faculties perfect, under the conditions to which they are exposed."†

Surely the most bigoted Intuitionist must admit that Nature, so far at least, is a Utilitarian; but Mr. Wallace avows his conviction that the theory of Natural Selection must be modified in its application to man in accordance with the facts of Mental Philosophy. Now should Mr. Wallace ever consider the discoveries made by Gall worthy of his attention, he will probably find that it is his idea of mental philosophy that requires modifying, and not the law of Natural Selection. Man is an exception to the rest of the animal creation, inasmuch as by his faculty of reason he changes Nature and adapts it to his requirements; but an animal having no such power of adaptation, is changed by Nature. But at the expense of a little more space let us give

* "Natural Selection," p. 302. † "Descent of Man," vol. 1, p. 98.

Mr. Wallace's own clear statement. He says: "From those infinitely remote ages, when the first rudiments of organic life appeared upon the earth, every plant and every animal has been subject to one great law of physical change. As the earth has gone through its grand cycles of geological, climatal, and organic progress, every form of life has been subject to its irresistible action, and has been continually, but imperceptibly moulded into such new shapes as would preserve their harmony with the ever-changing universe. No living thing could escape this law of its being; none (except, perhaps, the simplest, and most rudimentary organisms), could remain unchanged and live, amid the universal change around it. At length, however, there came into existence a being in whom that subtle force we term *mind*, became of greater importance than his mere bodily structure. Though with a naked and unprotected body, *this* gave him clothing against the varying inclemencies of the seasons. Though unable to compete with the deer in swiftness, or with the wild bull in strength, *this* gave him weapons with which to capture or overcome both. Though less capable than most other animals of living on the herbs and the fruits that unaided nature supplies, this wonderful faculty taught him to govern and direct nature to his own benefit, and make her produce food for him, when and where he pleased. From the moment when the first skin was used as a covering, when the first rude spear was formed to assist in the chase, when fire was first used to cook his food, when the first seed was sown or shoot planted, a grand revolution was effected in nature, a revolution which in all the previous ages of the earth's history had had no parallel, for a being had arisen who was no longer necessarily subject to change with the changing universe—a being who was in some degree superior to nature, inasmuch as he knew how to control and regulate her action, and could keep himself in harmony with her, not by a change in body, but by an advance of mind. * * *

Here, then, we see the true grandeur and dignity of man. * * * Man has not only escaped 'Natural Selection' himself, but he is actually able to take away some of that power from nature which before his appearance she universally exercised." * Natural Selection, however, has only been transferred from the body to the brain and mind; but this Mr. Wallace denies. He says: "Among civilised nations at the present day, it does not seem possible for natural selection to act in any way so as to secure the permanent advancement of morality and intelligence; for it is indisputably the mediocre, if not the low, both as regards morality and intelligence, who succeed best in life and multiply fastest. Yet there is undoubtedly an advance—on the whole a steady and a permanent one—both in the influence on public opinion of a high morality, and in the general desire for intellectual elevation; and as I cannot impute this in any way to 'survival of the fittest,' I am forced to conclude that it is due to the inherent progressive power of those glorious qualities which raise us so immeasurably above our fellow animals, and at the same time afford us the surest proof that there are other and higher existences than ourselves from whom these qualities may have been derived, and towards whom we may be ever tending." †

Now in what does man's great strength consist in comparison with other animals? As an individual naked savage he is probably inferior to several of the higher brutes. But man is not an individual only: by the ties of mind, the strongest of all ties, he is united through his social instincts, as the brutes are, with the family; and through his moral sentiments, which the brutes are not, with his race. It is with the great body of humanity, therefore, and not with individual man, that nature and brutes have to contend. It is this unity with his fellows that constitutes man's strength, and to this unity Justice, Courtesy, and Kindness, Faith, Hope,

* "On Natural Selection," p. 324. † Ibid., p. 330.

and Charity, the peculiar functions of the coronal region of the brain, are necessary. It is to his power of combination and co-operation that his great superiority may be traced, and "natural selection" is driving him in this direction as fast as the world can be prepared for him; and the superior minds, to which Mr. Wallace alludes, are the pioneers towards more perfect unity. In the meantime nature continues to secure the "survival of the fittest." In the age of war, when the most necessary thing was the protection of life and limb and property, the strong man—the warrior—was the fittest. We have passed on to the commercial age. What is now wanted is to conquer the earth, and to people it with the best breeds, and this is best done by unlimited competition, or free-trade, which means every one for himself and Devil take the hindmost; and the "fittest" for this is by no means the highest type of man—very nice feelings are only in the way. At present we have no more organisation than what fits us for this state of things, but with a more complete and perfect "Organisation of Industry," which Carlyle says is the question for this 19th century, we shall "secure the permanent advancement of morality and intelligence."

Mr. Wallace argues that the *practice* of virtues on the ground of their utility could not account for the *sanctity* which attaches to them even among savage tribes. Now "the sanctity attributed to moral distinctions" is owing to their being associated with religious feeling, in the same way that sanctity is given to the "fetish," and argues nothing in its favour.

Mr. Wallace, in his review of Darwin in the *Academy* (March 15, 1871,) says: "Mr. Darwin concludes that the moral sense is fundamentally identical with the social instincts, and has been developed for the general good of the community, rather than for its greatest happiness. 'General good' is defined as 'the means by which the greatest possible number of individuals can be reared in full vigour and health, with all their faculties perfect under the conditions to which they are exposed'; and it is quite conceivable that this may not be always identical with 'greatest happiness.' If so, the present theory will be a step in advance in the history of the utilitarian philosophy." I cannot, however, agree with Mr. Wallace. Just in proportion as such a theory was not "identical with the greatest happiness" would it be a step backwards instead of in advance. The greatest number of *individuals* may consist of savages or brutes, or creatures capable of even less enjoyment than they are. To rear the greatest number of them would certainly not be for the greatest happiness, and would therefore be a step backwards. To be a step in advance we must rear the greatest number capable of the greatest enjoyment.

Better fifty years of Europe than a cycle of Cathay.

"A moral sense fundamentally identical with the social instincts" would be just such a step backwards. The social instincts are possessed in full strength by the brutes, and would not raise us a single step above them. The social instincts—the love of wife, children, and friends—seek their aims at the expense of the community. The true moral sense, which alone has raised man, and is a step in advance, makes the social instincts—the interest of our own wife, children, and friends—secondary to the good of the community, to the general happiness. It is justice, courtesy, love of the race, and not of our own families, upon which the moral sense must be based.

It is "quite conceivable," however, that raising the lowest and most numerous class in society to their rightful position might not *at first* be identical with "greatest happiness;" but this would be only the common case under utilitarianism of sacrificing a present good to a greater in the future.

Again, Mr. Wallace tells us that the Brain of the Savage is larger than he needs it to be. The question of course is, what does he need his brain for? For the sensibility of which it is the organ—for his enjoyment, for feeling, and for as much intelligence as will *direct* his feelings towards their proper objects or enjoyments. Now the power of feeling is in proportion to the size of the brain, so that with increased size goes increased power of enjoyment. Mr. Wallace's error is the common one of those who neglect Gall's discoveries, and reject the knowledge his followers have attained of cerebral physiology and the nervous centres. He believes that the brain is the organ of the mind, but by mind he evidently means the intellect or intelligence. He confounds general size with special function: for the part of the brain connected with the intellect is a very small proportion of the whole, and bears in the savage a much nearer relation in size to the ape than to the philosopher. The phrenologist sees in the savage skull only indications of a more powerful animal—of power in the direction of the brute beast, with great destructiveness, courage, cunning, caution, and love of offspring such as peculiarly distinguishes the ape tribe; but of intellect he finds very little more than the brutes themselves possess. Both Mr. Wallace and Mr. Darwin would start from a much higher stand-point in their valuable researches if they would acquaint themselves with what has been already done by phrenologists in the same field. The comparative anatomy and physiology of the brain and nervous system, and of mental function in connection with structure, have been their study for 60 years and more, and most valuable atlases of plates have been published by Dr. Vimont, of Paris, and others. To me it is a most wonderful thing that to such men all this labour should have been labour in vain.

In reply to certain objections which I have heard against what is now often called Darwinism, I would observe that Instinct, which is organised and transmitted experience, is often more correct in its intimations, and goes straighter to its aim, than the first dawnings of reason among men. Reason is a generalising instinct, and in the first attempts of this new and untried faculty to guide experience and to attain its results at once, it often goes wrong; and as that wrong is fixed and made permanent by law and custom, or sanctified by religion, men, in some particulars, are worse off than the brutes. The instincts of many of the lower animals are far nobler than the habits of savage races of men, and this Mr. Darwin clearly shows.*

* Of course I am not unfamiliar with the objections that have been brought, and which may fairly be brought, against Mr. Darwin's theories. Still, Nature's plan seems to have been one of experiment; she has tried all ways, and retained the best. As Prof. Huxley says, "For the notion that every organism has been created as it is and launched straight at a purpose, Mr. Darwin substitutes the conception of something which may fairly be termed a method of trial and error. Organisms vary incessantly; of these variations the few meet with surrounding conditions which suit them and thrive; the many are unsuited and become extinct."—(Huxley's "Lay Sermons." p. 302.) I think it is impossible to deny this much, or that external conditions, with inheritance, reversion, natural selection, &c., play an important part. Still, as regards the transmutation of what, as yet, we call species, we cannot get over the general sterility of hybrids, and Prof. Macdonald is of opinion that "a due consideration of the progressive development of an embryo or germ within the Graafian vesicle militates against Darwinism in any attempt to press hybridism beyond the boundary of nearly allied species." It has also been truly said: "If species have developed through successive increments of minute variations, we ought to discover in the stone records of the earth's history very distinct evidence of transitional and intermediate forms. Not only so, but the number of such progressive steps having been innumerable, the intermediate forms should be discovered in preponderating numbers." "It is incredible," says Mivart, "that birds, bats, and pterodactyls should have left the remains they have, and yet not a simple relic be preserved, in any one instance, of any of these different forms of wing in their incipient and relatively imperfect functional condition."—("On the Genesis of Species.") On Specific Genesis, Mr. Mivart is inclined to attribute great power in the origination of

THE NATURAL LAW AND THE RIGHT TO LIVE.—A little boy, whose mamma had just blessed her husband with twins, was called in to admire his new relatives. He looked them carefully over, and then said, "I think we will keep this one." He had evidently heard of the drowning of supernumerary puppies or kittens.

"The Empress Elizabeth of Russia, during the war with Sweden, commanded the Hetman, or chief of the Cossacks, to come to Court on his way to the army in Finland. 'If the Emperor, your father,' said the Hetman, 'had taken my advice, your Majesty would not now have been annoyed by the Swedes.' 'What was your advice?' answered the Empress. 'To put the nobility to death, and transplant the people into Russia,' calmly replied the Cossack. 'But that,' the Empress observed, 'would be rather barbarous.' 'I do not see that,' said he; 'they are all dead now, and they would only have been dead if my advice had been taken.'"*

Both these ways of dealing with *surplus* population seem

new forms to an internal tendency, real, but as yet inexplicable—not, however, inexplicable to him who feels obliged to believe that

"There lives and works
A soul in all things, and that soul is God."

But whatever difficulties we may meet with, probably consequent on the imperfect state of our present knowledge of this subject, it is impossible to watch the progress by which mind is formed, and the gradual co-ordination of mind and body so as to work harmoniously together, without coming to the conviction that this laborious work of nature, taking almost infinite time, must have been transmitted from one form to another in the way at present familiar to us. Mind must have been perfected and transmitted to us through every variety of form, from the monad to the man; but how the gaps have been got over—how we have passed from one form to another—we do not at present know. But we are at the very commencement of the investigation.

* "Life of Sir James Mackintosh," vol. 2, p. 51.

reasonable enough; still they are not such as we are at liberty to follow. A farm labourer or weaver with 12s. a week, on getting a marriage license, thinks that *licence* enough to beget a dozen children, although he knows his own earnings cannot support one-half of them, and that his more thrifty and provident fellow-labourers must work harder every day to help to keep them; and yet we cannot drown half of them, which would seem natural. So of a savage race that is dying out before civilisation: we are not at liberty to kill them off at once, although Cossack philosophy might say it would be all the same 30 years hence.

A deputation of workmen, waiting upon one of our Prime Ministers, pleaded in extenuation of their petition that they had "a right to live." The Minister said "he did not see the necessity"; and such seems to be the opinion of a good many people at the present day. They think our Poor Law, that prevents every one from starving, a mistake, and opposed to Natural Law, which they consider to be that only those that can keep themselves ought to live. But although this is the law as it relates to other animals, it does not equally apply to man. He is an exception, as we have seen. With the brutes it is a struggle for life; with man only for that which will make life desirable, a bare subsistence being guaranteed to him. Man is a faggot of sticks, and it is not to his interest that the sticks should be parted and broken separately. His strength and happiness depend upon his unity, and this depends upon the full development of that part of his mental system—his moral sense—upon which this unity depends. It would give a most violent wrench to this system if any were allowed to starve while he could prevent it. It may be proved to be our interest, and therefore our duty, to do all we possibly can for our fellow-men; for by so doing only can we bring all our moral faculties into activity: by this means only can we exercise that part of our nature upon which our greatest happiness depends. "If a man will not work, neither shall he eat," says St. Paul; but as the German Professor said, "Ah, Paul! he was a clever man, but I do not agree with him." A man *unable* or *refusing* to work is the same in our view, the only difference between a physical and moral impediment being that the latter probably may be removed by a little gentle coercion when the former could not. We suffer in a variety of ways from the shortcomings and delinquencies of our fellow-men, and we may think it very unjust that we should do so, or that the industrious and provident should work to keep the idle and improvident; but it is not unjust, inasmuch as we gain in a thousand ways by the union with our fellows for one in which we lose. And if we suffer, it furnishes the most direct interest in the removal of all those impediments, physical, moral, and intellectual, to a more perfect union. The Social law, which is the Natural law, binds all together for good and for evil. We cannot oppress our fellow-creatures without injuring ourselves, or make slaves without enslaving our higher feelings. Before the abolition of slavery a worthy Alderman of London, an M.P., to prove the advantage of slavery to this country, told the House that it afforded a market for the refuse fish and corrupted food, which could be sold for no other description of persons.* This was certainly not taking the most enlightened view of the matter—only an aldermanic one. England in her commercial policy has still, however, too much of this spirit left as applied to the Slavery of nations. It is true we have a commercial age through which we must go before we can reach a better.

Nathaniel Hawthorne says: "It takes down the solitary pride of man, beyond most other things, to find the impracticability of flinging aside affections that have become irksome. The bands that were silken once are apt to become iron fetters when we desire to shake them off. Our souls, after all, are not our own. We convey a property in them to those with

* "Memoirs of the Life of Sir Samuel Romilly," vol. 2, p. 426.

whom we associate; but to what extent can never be known, until we feel the tug, the agony, of our abortive effort to resume an exclusive sway over ourselves."* We would not, however, give up all the pleasures of Friendship because our friends may be false or die, and the same applies to all our other human ties; and fortunately we are not often called upon "to resume an exclusive sway over ourselves." Here, as elsewhere, "all which we enjoy, and *a greater part* of what we suffer, are put in our own power."†

The causes of Pauperism are unlimited competition in the "struggle for existence," in which the weakest necessarily go to the wall; our Endowed Charities, including the Poor Law, by which the self-reliance, self-dependence, and self-respect of the poor are undermined; and the systematic breeding of paupers by the Poor Law authorities—for there is a class to whom the workhouse is an inheritance, the recipients of parish relief consisting principally, not only of the damaged and spoilt ones, but of the children of paupers to the third and fourth generation. The remedies are a free circulation of labour from where it is in excess to where it is wanted, in all parts of the empire; the absorption of our eleemosynary charities into Provident and Educational Institutions; and a more careful rearing and training of the rising generation of paupers; good food, good air, and good education being provided at whatever cost real *efficiency* requires. Recollect we have a deteriorated constitution, bodily and mental, to deal with, and the first loss to the community will be the least.

## THE PRACTICE OF MORALITY.

If, then, all reason and all facts point to the conclusion that the end and aim of creation is the production of the

* "The Blithedale Romance," p. 236. † "Butler's Analogy."

largest amount of enjoyment, the question is, how may that be best brought about? We have seen that man has no faculties that induce him directly to seek happiness: all his faculties have other objects, but happiness is the result when such objects are attained. He has certain wants, and has pleasure in gratifying them; and it is the aggregate of such pleasures that constitutes what we call happiness. The sum of enjoyment, then, will be in proportion to the gratification of all the wants, and it has hitherto been supposed to be the peculiar function of Morality to see that these wants are not gratified at the expense of other people. But this has been a narrow and restricted view of the subject, as the great object of Morality is to make all as happy as possible. "*Thou shalt* is written upon Life in characters as terrible as *thou shalt not*, although poor Dryasdust reads almost nothing but the latter hitherto." *

Morality has been defined as "the Science that teaches men to live together in the most happy manner possible." They have been tied up, like a bundle of sticks, with their fellows, and upon this unity their strength and happiness mainly depend, and what has especially been called Morality are the laws which best enable them to fulfil the objects of this unity, and move harmoniously with their fellows. A special sanctity has been given to these laws, as if they existed independently of man; but that has been a mistaken and shortsighted view of the matter, and we are beginning to learn now that the observance of all law is equally our duty and equally sacred. We have to organise such conditions as will best enable each creature to gratify all his wants and bring all his powers into active operation, and this consistently with the equal rights of others. We are here, however, necessarily mixing Moral and Social Science; for it is the duty of Moral Science to teach us what our duties are, and of Social Science to place us in the circumstances in which

* "Cromwell's Life and Letters," by T. Carlyle, p. 228.

they may be best performed. We have shown what our wants are, and these point directly to our duties: the faculties and attributes with which we have been endowed all show for what we have been made, and what therefore are our duties:—

> All declare
> For what the Eternal Maker has ordained
> The powers of Man.—*Akenside.*

We have to ascertain, then, what *are* the faculties, and what constitutes the proper use of them, and what their abuse. I have endeavoured to give a list of them, and therein have forestalled many of the remarks that properly come under this head. It will be impossible here to do more than lay down a few general principles.

Our object is to produce the largest amount of enjoyment. Obedience then is required to all the laws, physical, organic, and mental, as well as moral; for suffering follows equally a breach of any of them. We find a sound mind only in a sound body: obedience to the laws of health, then, is the first moral duty. Pious people live in ignorant disregard of all nature's laws, and thus necessarily bring misery on themselves and families; and because they go regularly to church or chapel, and pray a good deal, and keep the *ten* commandments, they talk of the mysterious dispensations of Providence. It does not matter how good a person may be in all other respects, if he puts to sea in a ship that is not seaworthy, he will go to the bottom; if he marries a sickly wife, however pious she may be, he will probably have sickly children; if the mind is uncultivated, and he is too ignorant to guide his wish to do right and all his other instincts, which are mere blind impulses, he will suffer accordingly, however moral he may be in the common acceptation of that term.

Our capacity for enjoyment is dependent upon our freshness. A jaded and over-used faculty is capable of little enjoyment. All pleasures pall upon too frequent repetition. This is well known with respect to all our Propensities, and even the Æsthetic feelings tire with too much use. A month's travel in a beautiful country, with attention directed only to its beauties, reduces the pleasure first to indifference and then even to a pain. This is why we should vary our occupations, so as daily to exercise all our wants and feelings, and call all our faculties into activity. If "one franc's worth of coal does the work of a labourer for twenty days," and work is properly directed, and its produce fairly divided, surely that ought to give time for every variety of mental and bodily labour and enjoyment. If we use one faculty, or class of faculties, too much, it must necessarily be at the expense of others, as there is only a certain amount of force available for the exercise of all. A man who pays exclusive attention to money-making, which he calls business, soon disqualifies himself for the use of all those higher feelings which peculiarly distinguish him as man. These feelings are delicate plants, and soon fade unless they receive their proper aliment and exercise: imperfect nutrition attends enforced rest, and atrophy follows. A man who puts off his enjoyment till he has made his fortune, finds that age has deprived him of many pleasures of which he was once capable, and that he has also sold the best part of his soul to the devil of acquisition. In fact, he has lost the end in the means. "Politics were never more corrupt and brutal; and Trade, the pride and darling of our ocean, that educator of nations, that benefactor in spite of itself, ends in shameful defaulting, bubble, and bankruptcy, all over the world. * * But the great equal, symmetrical brain, fed from a great heart, you shall not find."* This is why we rise so little above Dinners and Display, and the prudent effort to secure a good place in another world, which we call Religion. A man may make money, but he often limits rather than extends his

* Essay, "Work and Days," Emerson.

power of happiness in so doing. He may make a mountain of gold, but

> Can gold gain friendship? Impudence of hope!
> As well mere man an angel might beget.
> Love, and love only, is the loan for love.
> All like the purchase; few the price will pay;
> And this makes friends such miracles below.

If we depend upon the selfish feelings for our enjoyments they will fail us in our old age. It is the predominance of the unselfish feelings that is required, and this is only to be had by their daily exercise. The man who is habitually just, courteous, and kind, who creates all the happiness he can, is making an atmosphere around him of respect and love, and a bed which he will find easy as he grows old. Selfish enjoyment is an unit; to the unselfish is added the reflected enjoyment of all around. If to this is joined a cultivated love of the beautiful, and an acquired taste for the best reading, so as to place us *en rapport* with all the great and good that have lived before us, with all that is so wonderful in art and science in the world around, and with all that is high in modern thought as it arises, we have made a fortune for our old age, larger in amount than all the gold in the world, for all the gold in the world could not purchase these things. It is a too common and fatal mistake to suppose that it can, and it is also a fatal mistake to suppose that we can put off the acquisition of these aptitudes till we are old. It is the daily exercise of all our faculties that alone can give them. Sir Joshua Reynolds says "that a relish for the higher excellencies of art is an acquired taste which no men ever possessed without long cultivation and great labour and attention." It is habitual dispositions that we require to form. Virtue should be a habit, automatic in its action. We should do right without thinking about it. A mean, lying, untruthful, unjust, unkind, or cruel thing should be an impossibility to us; and it is only constantly repeated action of the highest feelings, where they are not naturally strong, that will induce this state.

Of course this kind of training is based on the supposition that there are certain general rules that experience has tested and shown to be right, and which we have admitted to be right, and from which, therefore, there should be no appeal. It will not do to be constantly calculating consequences in individual acts, or in what may be considered particular circumstances; for the heart is a very special pleader, and character, much more than is supposed, determines opinion. We must make up our minds on the best calculation that we individually are able to make, joined to the experience of all the great and good, and then the performance of duty, *i.e.*, doing on all occasions what we have thus previously determined to be right, regardless of present consequences, will be found to be the most direct and safest road to happiness.

There are many commandments or general rules about which there is no doubt, but there are many questions—the relations between man and wife, the relations between capital and labour, for instance—yet to be settled. "Thou shalt not steal." True; but is taking from the labourer an unfair share of that which capital and labour have jointly produced, because competition enables us to do it, stealing? Because, if it is, most of the large fortunes now in the world have been thus acquired. Is it right to lie that good may come, as the Jesuits do? No; because in the long run no good ever *can* come from it. If the principle were admitted, the trust and reliance on each other upon which society is founded could no longer exist; and this is infinitely more valuable than any good that could come from deceiving in special cases.

"To do to others as we would they should do to us," is far too general a rule for special guidance, and can only be acted upon in its broadest spirit. This moral axiom is based on the supposition that all characters, and therefore all wants, are alike; whereas, all characters differ, as much as a pig from an angel, and what would suit a pig to have done to it would not at all suit the higher nature. No doubt an alderman, when he is inviting a large company to dinner, is doing as he would others should do to him. "We are to do as we would be done by"—as the young gentleman said when he kissed the pretty girl—does not point to much self-sacrifice, and is certainly not indicative of the highest order of morality; and yet such an application of the rule is in the spirit of at least half the practice that would be brought under this law. A moral system in the present day requires something much more definite. Our own wants are a most imperfect standard of what would produce the largest amount of enjoyment to others.

The doctrine of non-resistance is opposed to the whole of nature's teaching. Nature resists and makes us suffer whenever we do wrong, with the direct object of teaching us we *are* doing wrong, and of making us do better in future. "To return good for evil," "to heap coals of fire on our enemies' heads by kindness," is again a question of character, and is very well when applied to the higher natures, but it is only a premium for ill-doing in the lower. Kindness and reason to an unjust but generous enemy make him repent at once. A man insults you, the object is to prevent any repetition of the act, not to revenge the past. Kindness and reason are the most direct way to this where you have to deal with a good man, but they would be thrown away upon a bad one, and the only way to prevent a repetition, probably, of the insult, would be to knock the latter down. If so, knock him down; he will abstain another time from fear, when he would not from any more generous motive. Love and kindness would not prevent aggression in a tiger, and there are many men with tiger natures.

We have seen that the object of pain or suffering is

reformation—to teach when we have done wrong that we must amend it. To relieve the suffering, without securing the amendment, is not always "going about doing good." The entire character and practice of our eleemosynary charity require modification in accordance with this law. Most of our charitable institutions are at present only premiums for improvidence; they make a man's well-being depend upon anything but his own efforts, guided by intelligence and character; they attack at the base—they cut up by the root—self-reliance, self-dependence, and self-respect, upon which all good character, all tendency to rise, is founded among the poorest classes. If a man once takes to charity-hunting—if, indeed, he is taught to depend upon anything but his own efforts, it is moral death to him and physical misery. In the cities where these eleemosynary charities abound, their pauperising and demoralising influence is incalculable. For three who trust to such charity only one succeeds; the effect on all is bad, but to the unsuccessful it is most disastrous. Their own efforts would have given them what they wanted in half the time they have spent in begging; but the greatest loss is in character. Only those charities, then, that help the poor to help themselves are really doing good.

Lecky tells us that "Magnanimity, self-reliance, dignity, independence, and, in a word, elevation of character, constitute the Roman idea of perfection;" while humility, obedience, gentleness, patience, resignation, are Christian Virtues. "The duty," said St. Jerome, "of a Monk, is not to teach, but to weep." "The business of the Hermit was to save his own soul." "The serenity of his devotion would be impaired by the discharge of the simplest duties to his family." * "A law of Charlemagne, and also a law of the Saxons, condemned to death any one who ate meat in Lent, unless the Priest was satisfied that it was a matter of absolute necessity" (p. 257). Among the Poles the teeth of the offending per-

* "European Morals," vol. 2, pp. 72, 156, 134, 143.

son were pulled out. When martyrdom was the reward of Christianity, in its early stages, none but the higher minds joined it; but when it became a State Religion it was joined by the world, the flesh, and the devil, and was lowered in its tone proportionally; so that, as Mr. Lecky tells us, the 1,000 years after Constantine were the most contemptible in history. There has been no other enduring civilisation so absolutely destitute of all the forms and elements of greatness, and none to which the epithet *mean* may be so emphatically applied (pp. 14, 72). To religious ignorance and bigotry, to "the doctrine that correct theological opinions are essential to salvation, and that theological error necessarily involves guilt," may be traced almost all the obstructions they have thrown in the way of human progress—adding immeasurably to the difficulties which every searcher after truth has to encounter, and diffusing far and wide intellectual timidity, disingenuousness, and hypocrisy (pp. 377, 420). So that "not till the education of Europe passed from the monasteries to the universities, not till Mahommedan science, and classical free-thought, and industrial independence broke the sceptre of the church, did the intellectual revival of Europe begin" (p. 219).

We are glad to find Mr. Lecky leaning so decidedly to our views. He says: "In the eyes both of the philanthropist and the philosopher, the greatest of all results to be expected in this, or, perhaps, any other field, are, I conceive, to be looked for in the study of the relations between our physical and our moral natures." This relation alone goes to the root of the matter, and the strongest force is always connected with the largest organs or combination of organs. Moral Science is a pure system of dynamics—the action of the Will always representing the strongest force. The Romans were nothing but a nation of robbers, the strongest of their time: their strength depending upon the perfection of their training and organisation. Patriotism was a necessary part of this organisation, making the good of one dependent on the good of all; and patriotism was their strongest virtue, and the foundation of all their others. They made war upon weaker nations, killed the men or made slaves of them, stole all their goods, and carried them with the women and children in *triumph* to Rome, till Rome became so rich from these spoils that luxury ultimately destroyed the nation itself. For ages among the Romans Combativeness and Destructiveness were the feelings in predominant activity—to fight and to destroy were their business, their daily occupation, so that at last, as we have before said, the main amusement of all classes were the gladiatorial shows, the spectacle of bloodshed, of death and torture. The heads of noble Romans all show predominating Destructiveness, with small Benevolence.

It is the same with our criminal classes. They are merely powerful or cunning animals, and with greatly predominating propensities nothing better ought to be expected of them, and they ought to be treated as such. At large, they are as certain to prey upon society as a tiger. "The relation between our physical and our moral nature" has long been evident enough to the Cerebral Physiologist. He has long known, beyond all doubt, that in proportion as the animal, moral, or intellectual region of the brain predominates, does the man or mere animal predominate in the character. If the intellect and moral region prevail, we have a moral man according to his lights. If the three regions are equally developed, the man will depend upon education and the circumstances in which he is placed; with the animal region decidedly in excess, we have a brutal animal; with the animal region and intellect predominating, a clever rogue. These various degrees of brain development are evident at once to the practised Phrenologist; and in 1836 Sir G. S. Mackenzie petitioned Lord Glenelg, then Secretary to the Colonies, that our knowledge on the subject might be used in the classification of criminals; of course, however, without effect.* In the state of public opinion, then as at present, the petition could not be granted. The conclusion we have arrived at 34 years later is no further advanced than that some kind of Intellectual test should be applied in the choice of applicants for the Civil Service, apparently however in ignorance of the fact that such an examination furnishes no test of *character* whatever. It gives a very imperfect indication of the quality of the instrument, but none whatever as

* "At present," he said, "our criminals are shipped off, and distributed to the settlers, without the least regard to their character or history." * * "There ought to be an officer qualified to investigate the history of convicts, and select them on phrenological principles. That such principles are the only secure grounds on which the treatment of convicts can be founded, proof may be demanded, and it is ready for production," etc. In a separate letter, Sir George said, "men of philosophical understanding and habits of investigation have been brought to perceive that a discovery of the true mental constitution of man has been made, and that it furnishes us with an all-powerful means to improve our race. * * Differences in talent, intelligence, and moral character, are now ascertained to to be the effects of differences in organisation. * * The differences of organisation are, as the certificates which accompany this show, sufficient to indicate *externally* general dispositions, as they are proportioned among one another. Hence, we have the means of estimating, with something like precision, the actual natural characters of convicts (as of all human beings), so that we may at once determine the means best adapted for their reformation; or discover their incapacity of improvement, and their being proper subjects of continual restraint, in order to prevent their further injuring society. * * And if, as thousands of the most talented men in Europe and America confidently anticipate, experience shall convince you, your Lordship will at once perceive a source from which prosperity and happiness will flow in abundance over all our possessions. In the hands of enlightened governors, phrenology will be an engine of unlimited improving power in perfecting human institutions, and bringing about universal good order, peace, prosperity, and happiness."

to how it may be used. Such an examination is as likely to furnish a clever rogue as an honest, persevering, good man. Of course it is better than none, as it keeps the fools out, if it lets the knaves in.

When these views are appreciated they will lead to the introduction of the same reforms into our Prisons that have, with much labour, been introduced into our Lunatic Asylums, and upon precisely the same principle—namely, that the criminal is not responsible, and must be reformed with as little suffering as possible; and where reform, as in very many cases, is impossible, he must be sorted and caged as other wild animals are at the Zoological Gardens. "It is society preparés crime, and the guilty are only the instruments by which it is executed."* Although Cerebral Physiology has almost stood still for the last 30 years, yet here and there we are not without evidence of a glimmering of light on this subject among physiologists. Thus Dr. Robert Bird says: "Men are great or little, good or bad, not by the influence of their schools and colleges, but in spite of them. Then is education of no avail at all, and are we to give it up? No, but let us give it its proper place and just weight in the general estimate. It is the oil to the wheels, and the varnish to the surface, but not the substance. It seems to me that we shall make little progress in the improvement of our race till we give our moral and mental philosophy a physiological foundation, instead of the metaphysical and sandy one upon which it now rests; till we judge and treat our brains as we now judge and treat our livers. There was a time when insanity was looked on as the work of a devil, and holy men were called on to exorcise him. Now, what should we think of a nation which believed and acted on such a doctrine in these days? We should pity it, and pronounce it plunged in barbarism and superstition.

* Quetelet.

But here we are asked what has vice to do with lunacy? What have they in common? My answer is, they are the same; they differ in degree, but they are the same in kind, and the sooner we admit and act on this, the better for ourselves. The phenomena of vice are as much the consequences of conditions of our tissues as are the phenomena of lunacy."*

* "Physiological Essays," p. 226.

## CHAPTER IV.

### PHYSICS AND METAPHYSICS.*

In metaphysics each disputant has his own language, and all controversy ought to begin by mastering each other's dictionary. Let us begin, however, at the very beginning, and see if modern science has thrown any light on this subject.

Hume says: "We may observe that it is universally allowed by philosophers, and is, besides, pretty obvious of itself, that nothing is ever really present with the mind but its perceptions, or impressions, and ideas, and that external objects become known to us only by those perceptions they occasion. Now, since nothing is ever present to the mind but perception, and since all ideas are derived from something antecedent to the mind, it follows that it is impossible for us so much as to conceive or form an idea of anything specifically different from ideas and impressions. Let us fix our ideas out of ourselves as much as possible; let us chase our imaginations to the heavens, or to the utmost limit of the universe, we never really advance a step beyond ourselves, nor can perceive any kind of existence but those perceptions which have appeared in that narrow compass."

This, however true, and, as Hume observes, now "universally allowed by philosophers," is not, however, true to common apprehension, which still believes in an external world, exactly as it appears, and knows nothing of consciousness, and that the objects of knowledge are in reality not things, but

* The substance of this Chapter has already appeared in the *Anthropological Review* for October, 1869.

ideas. But if, as Hume says, "external objects become known to us only by those perceptions they occasion", then we ought to be consistent in our inductions and deductions from this fact, and not attempt to raise whole systems on supposed knowledge beyond. We know nothing of Matter and Motion, of Spirit and Force; our own consciousness is all that is *known* to us, and all else is only more or less probable inference. Every external fact requires to be translated into the language of our thoughts; into our peculiar, and definite, and very limited "modes of thought." The longest chain of physical causation necessarily has its last link in the mind. The bridge to be built, or the road to be travelled, is not from physics to metaphysics, but from metaphysics to physics. The *ego* produces the *non-ego*. This, as I have said, is not the common apprehension: no doubt the first question is, What is the world without? The next, What am I? and thus on reflection we come to consider the medium or instrument by which the world becomes known to us, and which we call the mind, but which in reality is merely our consciousness. We ask, then, what is this consciousness? Whence comes it, or what is the cause of it? and lastly, What is the good of it—what is the use of it—what is the object of it? Here we have the questions of being or existence, of efficient cause or final cause, all questions of pure metaphysics, all requiring to be answered before physics can be properly pursued; for as we know of an external world only through the medium of our consciousness, how do we know that consciousness tells us truly, or to what extent its indications may be trusted?

We ask first, then, what is consciousness? It is a succession of varied feelings and ideas, and this only; differing, however, greatly in intensity. We call this variety of sentience by the names of sensation, propensity, sentiment, ideas, perception, conception, memory, imagination, and judgment. We speak of ideas and feelings passing through the mind, but there is no evidence of their passing through anything. The aggregate of these ideas and feelings *is the mind*, and there is nothing else. Consciousness is supposed to be a general term denoting states of mind, but mind has no existence in itself, but consists of these "states", or stream, or succession of thoughts and feelings. Consciousness, and sentience or feeling, in one sense are the same; but what is generally meant by the term is self-consciousness, which is the action of one faculty upon another—that is, reflection on consciousness. With Dr. T. Brown and James Mill, to have a feeling, and to be conscious of that feeling, are the same things; and this may be said to be the case with animals generally, who have feelings, but do not attend to them; but with J. S. Mill it is one thing to have a feeling, and another to recognise and reflect upon it, and refer it to one's-self, and to the series that make up our sentient existence. Self-consciousness probably requires another intuitional feeling besides reflection to be associated with the train of thought. It then induces us to refer all our states to the "I," or self, and is an element in our belief in personality.

But what do we *know* of consciousness? Being conscious and knowing are to us the same things. Consciousness in its several states of thought and feeling, of pleasure and pain, is the only real and absolute knowledge we have; all else is relative. Metaphysics we *know*; physics we only know in the relation to metaphysics, and as the facts and laws of physics are translated into ideas, the language of our consciousness. Phenomena and their laws are known to us but as parts of our consciousness. Much is said about observation and facts as opposed to mere thought in apparent forgetfulness of this truth, that every fact must become a thought before we can know it. All that may or may not be *without* ourselves, whatever properties such existences may possess, or whatever may be their real nature, all we know of them, or can know, is what appears in these thoughts, and although the space is very limited—almost as confined as our five senses—yet we cannot possibly go beyond.*

But whence comes this consciousness? What is the cause of it? The common answer is, that it is the action of the soul or of the mind; but we do not find either soul or mind in our consciousness, and that is all of which we have any knowledge; that is, if there is anything more than the suc-

* Prof. Huxley says: "It is conceivable that some powerful and malicious being may find his pleasure in deluding us, and in making us believe the thing which is not, every moment of our lives. What, then, is certain? What even, if such a being exists, is beyond the reach of his powers of delusion? Why, the fact that the thought—the present consciousness, exists. Our thoughts may be delusive, but they cannot be fictitious. As thoughts, they are real and existent, and the cleverest deceiver cannot make them otherwise. Thus thought is existence. More than that, so far as we are concerned, existence is thought, all our conceptions of existence being some kind or other of thought. * * The necessary outcome of Descartes's views is what may be termed Idealism; namely, the doctrine that, whatever the universe may be, all we can know of it is the picture presented to us by consciousness. This picture may be a true likeness—though how this can be is inconceivable; or it may have no more resemblance to its cause than one of Bach's fugues has to the person who is playing it; or than a piece of poetry has to the mouth and lips of a reciter. It is enough for all the practical purposes of human existence if we find that our trust in the representations of consciousness is verified by results; and that, by their help, we are enabled to walk surefootedly in this life. * * But it is also that Idealism which refuses to make any assertions, either positive or negative, as to what lies beyond consciousness. It accuses the subtle Berkeley of stepping beyond the limits of knowledge when he declared that a substance of matter does not exist; and of illogicality, for not seeing that the arguments which he supposed demolished the existence of matter were equally destructive to the existence of soul. And it equally refuses to listen to the jargon of more recent days about the Absolute, and all the other hypostatised adjectives, the initial letters of the names of which are generally printed in capital letters; just as you give a Grenadier a bearskin cap, to make him look more formidable than he is by nature."

cession of our consciousness, either soul, or mind, or spirit, consciousness does not recognise, and, indeed, has no knowledge of it. We find, however, a body, and that body has a brain, and pressure on that brain puts a stop to consciousness at once, and on further inquiry we find that whatever affects the brain affects our consciousness, and consequently we are obliged to come to the conclusion that there is a direct and immediate connection between them. Upon this philosophers are pretty well agreed, but as regards the nature of the connection there is at present very little agreement, if, indeed, there is any definite opinion at all. "If there is one thing clear about the progress of modern science," says Prof. Huxley, "it is the tendency to reduce all scientific problems, except those that are purely mathematical, to problems of molecular physics—that is to say, to the attractions, repulsions, motions, and co-ordination of the ultimate particles of matter. Social phenomena are the results of the interaction of the components of society, or men, with one another and the surrounding universe. But, in the language of physical science, which, by the nature of the case is materialistic, the actions of men, so far as they are recognisable by science, are the results of molecular changes in the matter of which they are composed; and, in the long run, these must come into the hands of the physicist."* Dr. Henry Maudsley tells us that "With every display of mental activity there is a correlative change or waste of nervous element; and on the condition of the material substratum must depend the degree and character of the manifested energy, or the mental phenomena."† And so also Professor Tyndall: "I can hardly imagine that any profound scientific thinker, who has reflected upon the subject, exists who would not admit the extreme probability of the hypothesis, that for

* "The Scientific Aspects of Positivism."—*Fortnightly Review*, June, 1869.

† "The Physiology and Psychology of Mind," p. 13.

every act of consciousness, whether in the domain of sense, of thought, or of emotion, a certain definite molecular condition is set up in the brain; that this relation of physics to consciousness is invariable, so that, given the state of the brain, the corresponding thought or feeling might be inferred; or given the thought or feeling, the corresponding state of the brain might be inferred." * From this acknowledged fact the erroneous conclusion is often drawn that thought is a phenomenon, or mode of action, of the brain; but although consciousness is connected with molecular motion in the brain, that motion does not constitute it; it is something separate and distinct, and may be exhibited apart from that motion. As Professor Tyndall says, "molecular forces determine the *form* only which the solar energy will assume;" and Dr. Bird also, "This is the chief function of all nervous centres, that they should in some way change the modes of motion communicated to them from without. And having conceived this, we begin to understand how the motion of the nerves of special sense are transmuted in the brain-tissues into those nervous conditions known as pain and pleasure, hope and joy, anger and love, memory, reasoning, and the play of fancy, &c." † When we say that thought is a function of the brain, we do not mean that it is the motion of the brain, but the power of the brain. Motion is nothing in itself, that is no entity; it is merely the transference of substance or entity from one point in space to another; it is the cause of this, or agent, which concerns us, and this we call Force.

To say that motion is a condition of matter, and that one body conditions, or sets in motion, another, is no explanation, for it still leaves unexplained *how* does it condition it? Let us take a simple illustration. A grindstone is set in motion, and we are told that the movement is something distinct from

* "Opening Address at Norwich," 1868.

† "Physiological Essays," p. 142.

the grindstone, and that there is nothing more real than "movement." Real,—but no-thing, *i.e.*, no entity; but how can a thing be real which is no-thing, *i.e.*, nothing? A moving grindstone is simply a grindstone moving, as distinguished from a grindstone at rest; it is the same grindstone in another or altered condition. Motion is mere change of place, and adds nothing, and takes nothing away from a body; but the cause of the motion does. When the grindstone was set in motion, something was added to it which is indestructible, and the motion is simply the indication of its presence. It is no part of the grindstone, and may be passed on, the grindstone simply determining the mode in which it is passed on. This something we call *force*. But whence was it derived? If the stone was set in motion by the hand, the force came from the food, and the force of the food from the sun; if by steam, from the coal, which again came from the sun; if from falling water, again from the sun, which lifted the water. This force is a measureable quantity; it is measured by its ability, or by what it can do. It is the same with the function, or power, or force of the brain. It is derived from the food, and after passing through various modifications which intensify it, it enters the brain, producing what is called molecular motion, and passing on as consciousness, as special thought and feeling, and each thought and feeling using up or expending an amount of force in proportion to its intensity. This thought or feeling is not a power of the brain, the force only passes through; the brain conditions it, or turns it into thought or feeling. Thus Herbert Spencer observes, "That no idea or feeling arises, save as a result of some physical force expended in producing it, is fast becoming a common-place of science; and whoever truly weighs the evidence will see that nothing but an overwhelming bias in favour of a preconceived theory can explain its non-acceptance;" and Dr. Henry Maudsley says, "Mind is the highest development of force, and to its existence all the lower natural forces are indispensably prerequisite.* Dr. Tyndall, however, says: "The passage from the physics of the brain to the corresponding facts of consciousness, is unthinkable." Why so? Of course that that which we believe to be the unconscious force of the brain can never think how it is it begins to think, is true enough; but in reality we have no such passage to make. Consciousness is all we know, or can know, and we cannot know, therefore, of anything differing from it; and we may reasonably object, therefore, to any argument based upon any supposed difference between mental and physical force, and to such terms as "physical" forces, and lower "natural forces", when they are made to imply a difference in *essence*, of which we can know nothing.†

Dr. Büchner says, "It is by the brain that we ascend from matter to mind." Now, the force that becomes conscious in the brain is not matter, but what is usually called spirit: matter does nothing but condition or transform force from unconscious to conscious. The proper way of expressing it would therefore be, it is by the brain that we ascend from automatic or unconscious mind to conscious. Surely this is thinkable, as conscious action is so frequently passing into unconscious. There is no instance in which we can be said to ascend from matter to mind. There is the transmitted vital spark, and the forces that bring it into activity. Dr. Tyndall truly says, "Given the state of the brain, the cor-

* "The Physiology and Pathology of the Mind," p. 67.

† Hume says: "We have no perfect idea of anything, but a perception. A substance is entirely different from a perception. We have, therefore, no idea of a substance. Inhesion in something is supposed to be requisite to support the existence of our perceptions. Nothing appears requisite to support the existence of our perceptions. We have, therefore, no idea of inhesion. What possibility, then, of answering the question, *Whether perceptions inhere in a material or immaterial substance*, when we do not so much as understand the question?"—"A Treatise of Human Nature," vol. i, p. 311.

responding thought or feeling may be inferred," and *vice versâ*. It is found that our varied consciousness of thought and feeling is connected with different parts of the brain, with intensity proportioned to the size and structure of the part, and to the extra force thereby employed. Let us take a less simple illustration of our meaning than the one I have given of the grindstone. Let us take a kaleidoscope. With each turn we get a new form. Now, what is the result? Let us never forget that force is indestructible; now, what force is there, and what has become of it? The force required for the turning of the instrument has probably passed into surrounding objects; "the form" taken by the glass in the instrument—the phenomenon, the appearance—the result of the action, passes away never to return, but the force that produced the form, the momentum underlying the phenomenon, is still in existence. Where? Certain waves of ether, or force, strike upon the glass, and producing a reaction of 477 millions of millions of waves upon the retina in a second, set the brain in motion, and produce the sensation which we call red; other parts of the glass produce 577 millions of millions in a second, producing the sensation of green, and 699 millions of millions produce violet, &c. Colour, then, is a sensation or idea; in the object it is a reactionary force, which passing through millions of millions of waves; its mode of action is changed by a nervous centre, and it becomes a sensation of red, green, or violet. But what then becomes of it? It is still a force, and a force that can produce 699 millions of millions of wave motions in a second must be a very strong one. This force, then, first sets other portions of the brain in motion, which give us ideas of form, size, position, &c, for we cannot think of colour apart from extension and figure, or of more than one figure without the idea of relative position, and it is from this mental state we derive our idea of space; and in this way a few simple forces received through our five senses from without, by the aid of the cunning machinery of the brain, are created into a whole world which we rather foolishly believe to have a real existence independently of the sensations, ideas, and feelings of which it is composed. This sensation of colour, as wo perceive, is within us, and not attached to the object, as most people believe; it is correlated or transformed force, reflected from, and modified by the something without or object; some objects in which we can perceive no more difference than we perceive in different bits of glass producing some millions of millions of vibrations more than others. As we started by affirming, the sensation, as part of our consciousness, is all we know, and, as Hume says, "we can never really advance a step beyond," for what possible resemblance can a sensation of colour have to the motion of the brain, or to the waves of ether, or to the reflected force from the object that caused those waves, and which is the only relation between the object and the sensation? It is impossible, then, that in any sense we can know things themselves, for we never come near them; there is always a long chain of antecedent and consequent between them and the consciousness they cause; the action of the brain, the action of the sense, the action of the atmosphere or the ether upon that, and the action of the object upon them; and how, therefore, the object affects the atmosphere or ether, and through them us, is all we know, or can know: *i.e.*, all the knowledge we can attain of things without our consciousness, of the *non-ego*, is simply how they affect, not ourselves even, but other things a long way off. Of course there is a world without us, but the world in which we believe is created by a correlation of forces in the brain, which forces are received in different quantities, and are variously modified before they reach the brain; and the study of these quantities and modifications, which we call phenomena and their laws, and which are known to us only as they are further changed and form part of our consciousness, is science. Mind, then, is correlated force: a few simple impressions are received through the senses, which are worked up by the complicated machinery of the brain into what we intuitively believe to be the world without us. A real world without consciousness would be the same as no world at all; and the world existing only in our consciousness is quite as good as if it had objective reality, and requires quite as much wisdom and power in its creation.

But what is Force? It is time we answered that question. Force is everything. It is known to us only as the ability or power to produce certain changes; but by force I mean the entity to which this ability or power belongs, from which it is no more separable than motion from the thing moving. It is not motion, but the cause of motion. It is not the action, but the agent. Deceived by appearance, we erroneously suppose force to result from the action of matter, whereas the action of matter in all cases is the result of force. We think the power is in the motion, but motion is merely the sign of the presence of the force. In the instance we have given, the grindstone, its working force seems to be in the motion, but the free force has been transmitted to the grindstone, and as soon as that has been used up—that is, has passed on—its grinding power and motion cease. To say in this case the force results from the action of matter, is a delusion of the senses—a vulgar error. It is altogether illogical; it is confounding the entity with its mode of action—the cause with the effect, the phenomenon with the noumenon, the motion with the thing moving. It is equally an error to confound the action or molecular motion of the brain with thought. The brain, like the grindstone, simply passes the force, or mental power, on, in the peculiar condition known to us as consciousness; the force is not the creation or the result of the action of the brain; it existed before it entered the brain, and it exists afterwards. We are told that "to suppose mental phenomena to be anything but phenomena is nonsense;" the nonsense must be in the supposition that phenomena are separable from noumena, or motion from the thing moving. We are told that "force viewed separately from matter is nothing:" this is merely a repetition of the same error, and is only true in one sense, viz., that force cannot be viewed separately from itself, for matter is force. "The phenomenon called matter arises from two forces, the one acting towards, and the other from a point or centre."* The existence of matter is based upon its supposed extension and solidity, but extension is an idea inseparable, as we have seen, from the sensation of colour, and solidity is in the same way based upon the sense of feeling, and that is mere repulsion—resistance to motion. J. S. Mill says, "when the question arises whether something that affects our senses in a peculiar way is, or is not matter, what seems always to be meant is, does it offer any, however trifling, resistance to motion?" In fact, solidity is a mere mental hypothesis invented in order to explain resistance, which is force. "We never touch matter even if it exist, and that we never see it is admitted alike by physiologist and metaphysician, for vision is merely a mental affection, called up by an impulse on the optic nerve, made by the movement of the luminous ether, which not the chair or table, but the force existing and acting external to the chair or table, or other object, radiates off."† Huxley says, "every form is force visible; a form of rest is a balance of forces; a form undergoing change is the predominance of one over others."

Science now reduces all things to "the attractions, repulsions, motions, and co-ordination of the ultimate particles of matter;" but these ultimate particles of matter—molecules and atoms, are the unknown quantities, the $x$, $y$, of certain forces, or centres of force—they are creatures of the imagination, and as pure assumptions as the spirits of the spi-

* "The Philosophy of Theism," p. 73.

† "On the World as a Dynamical and Immaterial World," by R. S. Wyld.

ritualists. If force viewed separately from what we call matter is nothing, what is it fills the interstices between these supposed atoms? Sir Isaac Newton supposed that all these atoms in our globe, if they could be brought together, would not occupy the space of a square inch; what, then, is the rest? in fact, what is compressibility? What draws the needle to the magnet? What fills the space between star and star, and draws each to the other? What draws the waters of the ocean to the moon? Is it the moon, at 240,000 miles away, or something that really exists between and unites the two? It is true that force may not be evident to our senses, except through its connection with what we call matter; but we mean by force the principle of action, the *unseen cause* of all change or motion, and as everything is in motion, so force is the universal interpenetrating medium throughout the universe. The forces at present known to us we call heat, light, electricity, galvanism, chemical affinity, attraction, and repulsion,—these, in their modes of action and manifestations, produce half the phenomena around us; the forces of life and mind produce the other half. These forces are indestructible and persistent, and as they pass readily into each other, we reasonably infer that they are in reality only one and the same force. Force is divided into active and passive. The active has been called spirit, the passive matter. The latter never originates, or is the cause of motion, but it determines the *mode* of motion—it "conditions" the active force, that is, regulates the manner in which it is manifested. This correlation, or transmutation, or change of force, takes place by the effect upon it of the medium through which it passes. The medium we call matter, and it is the "form" or visible force, or form of rest, or balance of forces before mentioned. Force is supposed to be produced by this matter or medium, but the "persistence of force" shows that this is an error; it is changed or conditioned only by the solidified or concentrated medium through which it passes. Force, and cause, and the ability to produce a change, are the same thing: but a cause implies two things—power or ability, and direction or determination; for every change or beginning has a purpose, a direction towards a definite object, and the force then includes both power and intelligence. This is not necessarily conscious intelligence. It may very reasonably be asked, what is *unconscious* intelligence? In automatic action, which is voluntary conscious action often repeated, we have the power, and all the effects of intelligence, where consciousness has ceased to attend it. We know nothing of mind in its essence, and can judge only by analogy as to its mode of action; and if, therefore, we may judge of universal mind by the small glimmer that appears in our own consciousness, we find conscious action constantly passing by repetition into unconscious; what at first requires a separate and distinct volition, goes on now without. Secondary Cause, or Natural Law, is thus simply automatic action of Universal Mind. That which originally required a distinct *conscious* volition, has passed, in the ages, into the fixed order of nature. We have thus intelligent *action* without consciousness or volition. *

* There is a class of philosophers who think that when they say a thing takes place "According to Law" they have explained the whole matter. Now "according to Law" means only the definite and unvarying *order* in which phenomena occur. The "Laws of Nature" are not the powers of Nature. Nothing is accomplished by them; they merely indicate the regulated way in which Nature works. Law tells us nothing of either cause—except as immediate antecedent—or direction, or purpose. Yet a whole school is content to stop there, and pronounce any further knowledge not only useless, but unattainable. Cause in Science is this order—this series merely of antecedent and consequent; but Cause in Metaphysics is the origin of this series or order. To say we know only phenomena is not correct. Consciousness is all we really know. We *infer* that consciousness is the result of molecular motion in the brain, but we also justly infer that there cannot be motion without something moved; and if that is a just

There is no bridge, then, from physics to metaphysics,—there is no road that way; the only road is from metaphysics to physics; for all force is mental force, such "will-power" as we are conscious of exercising in our small individuality. Universal Mind passes into unconscious action in general law, and again becomes conscious in the brain of percipient beings, in brains so modified that each sentient existence has its own special world, created by the correlation of the same external force in its own brain and nervous system. "The old specu-

inference we may with the same truth infer that there must be force, power or ability causing motion, and that as motion is inseparable from the thing moved, so is force equally inseparable from that of which it is the force. To say that we know *only* phenomena, is to infer the existence of motion without anything moving; to affirm the existence of force *only* as the cause of motion, is to separate the force from that of which it is the force, and it is as impossible to do that as to separate motion from the something moved. What feels—what is conscious? Not the brain, or the motion of the brain, but that which is passing through the brain which we call force, and which is the force of some entity unknown, and which must probably for ever remain unknown. But if we know nothing of its essence, we know something of its attributes. Now if we may go beyond Consciousness at all, and infer Motion, and say therefore we *know* phenomena, it seems to me to be equally reasonable to infer the cause of motion, and also to infer that *originally* there was a reason (beyond the line of least resistance) why it took place in one direction rather than another—that is, that there was purpose or aim or object in it. If there is no design or purpose in creation, then is there no guarantee for the stability of present Law and Order; if all has come to pass by chance, without what are called the "inherent" qualites of matter having received a difinite direction, then one more turn of the screw, one more change in the ceaseless motion, might take us back to Chaos. If there is purpose, there must be mind and intelligence, and probably power to enforce that purpose: and so *in fact*, wherever we find Force it is Force acting towards a special or definite purpose; it is therefore acting intelligently, and is mind automatic or conscious. In the universe all is motion or phenomenal, which motion is not chaotic, but acts in accordance with unvarying laws, which laws are everywhere combined to produce particular objects or results.

lations of philosophy, which cut the ground from materialism by showing how little we know of matter, are now being daily reinforced by the subtle analysis of the physiologist, the chemist, and the electrician. Under that analysis, matter dissolves and disappears, surviving only as the phenomena of force; which again is seen converging along all its lines to some common centre, sloping through darkness up to God."*

Berkeley denied the existence of an external world, and declared that our perceptions were mere ideas evoked in the mind by Deity; and we find that each brain creates its own world by the correlation of force from without, and as its conciousness is modified by its own structure. But force or power without cannot be separated from that of which it is the force or power,—that is, of Deity. Dr. Thomas Brown holds that the entity that has this ability *is* this ability; and Spinoza says, "God's power is the same as his essence"; and thus all manifestations are manifestations of the one Supreme Power, and all change "the varied God." Berkeley, in denying the existence of an external world, did not deny that there was *nothing* external; he admitted God's power, or mind, both conscious and automatic.

"Add intelligence to Force, and you arrive at conscious Force. What else is this but Will? What right have you to deny that wherever you find Force, there also is intelligence?"—"Higher Law," vol. 2, p. 164, by the Author of the "Pilgrim and the Shrine."

We separate the organic from the inorganic, the vegetable from the animal, &c.; but looking deeper we find all these individualities growing out of one root and moving towards one common purpose—life and enjoyment. Nature knows nothing of these distinctions made by man. Everywhere and in every part we are connected with the forces around us; growth and consciousness are the simple play in and out of us of these forces; if we could be cut off from them our individuality would collapse at once, life and feeling would cease. Our absorption of nature, and identity with the great whole, is perpetual. Our growth and the remotest star are inseparable. One unity underlies the whole.

* "The Reign of Law." By the Duke of Argyll.

Let us now consider what the bearing of these principles is upon certain physico-metaphysical questions of long standing. "Locke had shown that all our knowledge was dependent upon experience. Berkeley had shown that we had *no* experience of an external world independent of perception; nor could we have any such experience. He pronounced matter to be a figment. Hume took up the line where Berkeley had cast it, and flung it once more into the deep sea, endeavouring to fathom the mysteries of being. Probing deeper in the direction Berkeley had taken, he found not only was matter a figment, mind was a figment also. If the occult substratum, which men had inferred to explain material phenomena, could be denied, because not founded on experience; so also, said Hume, must we deny the occult substratum (mind) which men have inferred to explain mental phenomena. * * Matter is but a collection of impressions. Mind is but a succession of impressions and ideas."* This was felt to be unanswerable; but then it was not to be judged by its truth, but by its supposed consequences. It led to scepticism, whatever that may be, and the efforts of almost all metaphysicians since have been directed to the evasion of this truth. Lewes says: "Remark, also, that Hume's scepticism, though it reduces philosophy to a singular dilemma, viz., that of either refuting his sceptical arguments, or of declaring itself and its pretensions to be vain and baseless, nevertheless affects in no other way the ordinary judgments or actions of mankind."† Now, the only dilemma to which philosophy was reduced was, that it had found a truth that it did not then know how to use; but if, instead of being frightened, we had allowed it "to affect our ordinary judgments and actions", it would have shown that as we know only consciousness, the assumed differences between spirit and what has been called *gross* matter are pure assumptions, as are also all the arguments and superstitions that have been based upon these groundless assumptions,

* Lewes's "Biographical History of Philosophy," p. 491. † Ibid.

and it would have placed mental science on the same firm basis of induction as physical. We now know that "no idea or feeling arises, save as the result of some physical force expended in producing it"; that each idea and feeling is a separate correlation of force, and that the mind is merely, as Hume says, a succession of those ideas and feelings. There is but one entity in the universe, which Spinoza called Substance and physicists call Force. Mind is force, matter is force; consciousness tells us nothing of their essence, but they are the same in their manifestations and obey the same laws. The strongest force prevails equally in mind as in matter. The will is but the trigger that lets off this force in the direction of the object aimed at. The strength of the mental force is in proportion to the size of the organ through which it passes on its road to consciousness, and whose molecular motion induces that consciousness.

But mind is not a mode of action or motion—it is an entity itself; it is so much force, it is indestructible; it may change its form, but cannot cease to be. Each thought is a separate and indestructible existence. Mind, however, has no existence apart from these thoughts and feelings. It is not mind, then, that is indestructible, but these thoughts and feelings, or rather the force of which they are composed. If they cease to exist as thought and feeling, consciousness has again become automatic, and the force has passed on into some other form. The aggregate of these thoughts and feelings form the general or universal mind, and the universal mind is the only Individual Mind. Hume says: "'Tis still true, that every distinct perception, which enters into the composition of the mind, is a distinct existence, and is different, and distinguishable, and separable from every other perception, either contemporary or successive." The question is, what becomes of those "distinct existences"? Our own ideas of the world—our thoughts and feelings—are supposed to exist only in the percipient, in the "ego," but the force of which the ideas are composed existed separately from the "ego," and may, indeed must, do so again. That this mental, or voluntary, or will power, is in great part again converted into physical force, as heat and muscular motion, there can be no doubt. We have, as Mr. R. S. Wyld says, "a direct experience that the amount of physical power obtained is, *cæteris paribus*, in constant proportion to the amount of the mental effort which we are conscious of exerting in producing this physical power."* Certain narcotics and stimulants act directly upon the mind, greatly intensifying its action, and these immediately show themselves again in the expression of the eye, and in every muscle of the body. Tobacco suspends mental activity; opium and hashish greatly increase it. The venom of a snake that is sluggish, and only just awake, takes half an hour to prove fatal; but when excited and made angry its venom kills instantly. Illustrations of the direct connection between mental and physical force are everywhere around us, only little attention has yet been called to them. The torpedo strikes its prey at a distance of some feet. Here is mental force, or voluntary will power, acting at a distance through the medium of the water; and Prof. Owen tells us that the amœba draws its prey towards it as rubbed amber draws light substances. The mesmerist by voluntary effort can act upon others at great distances, through what medium he does not at present know, or whether any medium is required other than the all-pervading force around. Still the force that passes into consciousness in the brain is at present by no means sufficiently accounted for as to what afterwards becomes of it. I know that it is the orthodox opinion among men of science that "thought cannot exist without brain;" but this proposition, I hold, is by no means proved. Thought is not a motion of the brain, or a mere mode of action at all, for "it is force

* Paper read before the Royal Society of Edinburgh, "On Free, Anatomic, or Transmissible Power," May 3rd, 1869.

which causes the movement, and not the movement which causes the force;" and it must exist as force after the movement of the brain ceases, and why not as mental or conscious force? And if so, the question is what becomes of it? This inquiry becomes deeply interesting when at the present time we have many scientific men declaring that they recognise, in so called spiritual manifestations, both power and intelligence apart from brain.* A perception or idea is a positive

* I have seen very little myself of these so-called Spiritual Manifestations that I consider to be reliable. I have seen a man take my umbrella and get raps attended with extraordinary vibrations from the looking-glass, the door, the table and the floor; when he tried to get the same through the medium of a small walking stick, he failed till he took another person's hand. As the raps were always three, the number expected, it showed, I think, that the force, whatever it was, was volitional, or under the power of the mind. I have seen a table rise from the ground and settle again like a feather, no one touching it. I have seen people cured of nervous pains and affections by another's *silent* will at the distance of a large drawing room. I have also seen all the wonders of electro-biology, in which Dr. H. Maudsley tells us "the mind of the patient is possessed with ideas which the operator suggests, so that his body becomes an automatic machine, set in motion by them." I have seen also the organs of the brain brought one after the other into activity, by mere touch on the surface, under circumstances in which suggestion or thought reading was impossible. Mr. H. G. Atkinson has employed this method of Phreno-mesmerism to make many important discoveries in the functions of the brain which could not be made by Gall's method of comparing function with structure, the structure of the base of the brain and some other parts being but very imperfectly observed from without. This method, like anatomy and pathology, may supplement Gall's, but we must know more of mesmerism before it can be made equally reliable. For the discoveries made by Mr. Atkinson I must refer the reader to Chapter 10 of his work on "Man's Nature and Development." I have seen also, as I have said, thought-reading in which a mesmerised person who could not see, yet described all that was seen by another—the mesmeriser. The numbers of three watches, consisting of five figures each, were told consecutively; and probably precisely the same action was produced at the same time in both brains while thus *en-rapport*. The concentrated force emanating from the

entity; and again, I say, what becomes of all those ideas that pass through the brain? Do they retain consciousness, or merely possess the power to create the same ideas in others? Are they packed up somewhere in the brain for

brain, both conscious and unconscious, is sufficient to account for much, if not all, that I have seen. The curative power is probably the transmission of nervous force from one person to another, the pains of rheumatism being caused by the deficiency of such force. If two people sleep together, the one most deficient in vital and nervous force always draws from the other. This was clearly understood in King David's time, although the rationale might not be equally clear. I have also seen many other wonders produced by fruit, flower, seeing, and other mediums, and the only way I can account for them is the supposition that the love of deceiving amounts to a mania in some women, the same as kleptomania and other idiosyncracies manifest themselves in others. I should pronounce "Spiritualism" to be a nervous epidemic, based on powers of the mind at present unknown. Dr. Maudsley says: "It is most certain that there may be a reaction outwards of an ideational nerve-cell, independently of volition, and even without consciousness."—("The Physiology and Pathology of the Mind," p. 126.) We may expect fresh light to be thrown upon this subject every day, now that attention has been called to this intimate connection between Psychical and Physical Nature.

Mr. W. Crookes, F.R.S., has called attention to this subject in the *Quarterly Journal of Science*, July, 1870. He says: "A man may be a true scientific man, and yet agree with Prof. De Morgan, when he says: 'I have both seen and heard, in a manner which would make unbelief impossible, things called spiritual, which cannot be taken by a rational being to be capable of explanation by imposture, coincidence, or mistake. So far I feel the ground firm under me; but when it comes to what is the cause of these phenomena, I find I cannot adopt any explanation which has yet been suggested. * * The physical explanations which I have seen are easy, but miserably insufficient. The spiritual hypothesis is sufficient, but ponderously difficult.' Regarding the sufficiency of the explanation, I am not able to speak. That certain physical phenomena, such as the movement of material substances, and the production of sounds resembling electric discharges, occur under circumstances in which they cannot be explained by any physical law at present known, is a fact of which I am as certain as I am of the most elementary fact in chemistry.

future use? or have we souls made of them? As the body is made of protoplasms, or infinitely small vital cells, so may our spiritual body be made up of these forces called thoughts and feelings.

My whole scientific education has been one long lesson in exactness of observation, and I wish it to be distinctly understood that this firm conviction is the result of most careful investigation. But I cannot, at present, hazard even the most vague hypothesis as to the cause of the phenomena. Hitherto I have seen nothing to convince me of the truth of the 'spiritual' theory. I confess that the reasoning of some Spiritualists would almost seem to justify Faraday's severe statement—that many dogs have the power of coming to much more logical conclusions. Their speculations utterly ignore all theories of force being only a form of molecular motion, and they speak of Force, Matter, and Spirit, as three distinct entities, each capable of existing without the others; although they sometimes admit that they are mutually convertible."

Here Mr. Crookes, like other physicists, confounds "a form of motion" with its cause, the sign or symbol with the thing signified, a condition with an entity, and it is only dogs be thinks who would not! I am glad, however, that he has taken the subject in hand. Mesmer, like Gall, was a physician of Vienna, and his inaugural address was on the Universal Force, and Astral Influences on the Temperament and Destinies of Mankind: he must have believed in something else than mode of motion. The attitude of men of science towards this subject is anything but scientific. They ought to know that the study of the irregularities—and here are irregularities sufficiently well attested—leads to the knowledge of the true law. It was by the *irregularities* of the pendulum that the figure of the earth was determined; and it was by what first seemed to be *casualities* that polarised light was discovered; certain *irregularities* in the motion of Saturn and Herschel led to the discovery of the planet Neptune by Leverrier and Adams. "By mathematical reasoning and analysis they were enabled ultimately to point their telescope to that part of the heavens where the disturbing body ought to exist—*where it did actually exist*, and so to extend the knowledge of our planetary system twice as far into space as before," (*Journal of Psychological Medicine*, New Series, No. 2, April, p. 240); and if any physiologist or psychologist can but point his telescope in the right direction of these "spiritual phenomena," it may be of much more

I am told that Mr. C. F. Varley, the electrician, who sees ghosts, thinks that all the thoughts of our lives make up the body of the spirits. If a single drop of water may contain five hundred millions of animalcules—living creatures, and

importance than the discovery of the planet Neptune, and may extend our knowledge of mind, and of the conditions of happiness, twice as far as before.

Since this was written Mr. Crookes, as he promised, has been investigating, and he has given the results to the world in an article in *The Quarterly Journal of Science* for July, 1871, of which he is Editor, called "Experimental Investigation of a New Force." He has had sittings with Mr. Home, and by the aid of a "delicately poised balance" he has shown that Mr. Home was able, by merely resting the tips of his fingers on the table, to exercise a greater force than he himself could with his whole body; and by putting an accordion in a cage constructed for the purpose he has shown that it was played "without direct human intervention, under conditions rendering contact or connection with the keys impossible." Dr. Huggins, the astronomer, and Mr. Serjeant Cox were present, and bear separate testimony to the truth, fidelity, and accuracy of the experiments. Serjeant Cox says: "The results appear to me conclusively to establish the important fact, that there is a force proceeding from the nerve-system capable of imparting motion and weight to solid bodies within the sphere of its influence. I noticed that the force was exhibited in tremulous pulsations, and not in the form of steady continuous pressure, the indicator moving and falling incessantly throughout the experiment. This fact seems to me of great significance, as tending to confirm the opinion that assigns its source to the nerve organisation, and it goes far to establish Dr. Richardson's important discovery of a nerve atmosphere of various intensity enveloping the human structure. Your experiments completely confirm the conclusion at which the Investigation Committee of the Dialectical Society arrived, after more than forty meetings for trial and test. Allow me to add that I can find no evidence even tending to prove that this force is other than a force proceeding from, or directly dependent upon, the human organisation, and therefore, like all other forces of nature, wholly within the province of that strictly scientific investigation to which you have been the first to subject it. Psychology is a branch of science as yet almost entirely unexplored, and to the neglect of it is probably to be attributed the seemingly strange fact that the exist-if these "have organs of locomotion, the mode of which leaves no doubt that they each possess sensation and will, and that consequently they must have organs and tissues accordingly," there must be plenty of room to pack any amount of *ideas* in the brain.

Prof. Owen, in the last volume "On the Anatomy of Vertebrates," summing up his general conclusions, says: "Life is a sound expressing the sum of living phenomena. * * These phenomena are modes of force. * * Into these modes of force other modes of force have passed from potential to active states." Nerve force and electric force are convertible; and, asks the Professor, "may there not be conversion of force, magnetic, electric, thermotic, nervous? Is not thought to the brain what electricity is to the battery? Nerve force rises from reflex acts to volitional acts: the quiver of the pricked muscle to the desire to write a book. With the size and complexity of the brain centres, from Aztecs to Europeans, correspond the intellect and will." The Professor, subsequently, writing about what he calls "thought-force," says, "if lines of thought-force were visible, the ghost (of Samuel) would not therefore be more material." May, then, thought-force ever become visible? It is evident Prof. Owen does not think

ence of this nerve-force should have so long remained untested, unexamined, and almost unrecognised. Now that it is proved by mechanical tests to be a fact in nature (and if a fact it is impossible to exaggerate its importance to physiology and the light it must throw upon the obscure laws of life, of mind, and the science of medicine), it cannot fail to command the immediate and most earnest examination and discussion by physiologists and by all who take an interest in that knowledge of 'man' which has been truly termed 'the noblest study of mankind.' To avoid the appearance of any foregone conclusion, I would recommend the adoption of some appropriate name, and I venture to suggest that the force be termed the 'psychic force;' the persons in whom it is manifested in extraordinary power 'psychics;' and the science relating to it 'psychism,' as being a branch of psychology."

it impossible. Does what is called by spiritualists "a medium" supply the conditions? *

If consciousness or thought, as Prof. Owen says, "is to the brain what electricity is to the battery," no one believes that electricity ceases to exist when discharged from the battery, and yet the positive school of physicists always reason as if thought being discharged ceased to exist. Thus Dr. Louis Büchner, in "Matter and Force," says, "The brain is, then, only the carrier and the source, or rather the *sole cause* of the spirit, or thought; but not the organ that secretes it. It produces something which is not materially permanent, but which consumes itself in the moment of its production." How the brain can be merely the carrier of that of which it is the source or sole cause, or how *something consumes itself*, is not clear, but this passage furnishes a pretty fair sample of the Doctor's ordinary style of reasoning on Matter and Force.

The force that works the body is derived from the food, and is originally derived from the sun, in the divorce of the carbon from the oxygen in plants; and as "nerve-force and electric force are convertible," no doubt we are sensibly alive to the electric states around us. Temperature does not affect us as it does the thermometer, but according to the electric condition of the atmosphere. We often *feel* colder with the

* A Mr. Munder, of Boston, now of 630, Broadway, New York, was put upon his trial last spring for obtaining money upon the false pretence of taking spirit photographs. The trial lasted four days and he was acquitted, as his brother photographers, after every facility offered to them for investigation, had failed to detect any imposition. The *New York Sun*, of February 26, 1869, reports the case at considerable length. The reporter to this paper says he watched the process from beginning to end. After various sittings, with various results, always taking care that the glass was well cleaned and polished, "he thought he would try the effect of calling to mind the appearance of his father, as he looked just before he died, some eleven years ago. This time the negative gave a face in profile rather dim, but in general outline, he must confess, very like his father, as he thought of him."

thermometer at 50 deg. than at 40 deg. The *Spectator* (July 3, 1869), speaking of the intimate relation between the sun's spots and perturbations and the magnetic currents of our own earth, says: "Could we really establish any periodic law of electric excitement on the earth, it would not be irrational, but in the highest degree rational, to expect marked human phenomena in connection with it—either a great concurrent depression or a great concurrent stimulus to the energies of the human brain. * * In point of fact, it is by no means impossible that the issues of peace or war, of a financial crisis, or a religious agitation, may be closely bound up with these phenomena."

The force supplied to our organism is a definite quantity, and it has to work the whole machine. People who spend much time out of doors are ordinarily not great thinkers, and thinking and digesting cannot go on actively together; but the most important vital process is the putting in the new material to replace the old and worn-out, and although I suppose assimilation and decay are always going on, yet they go on most actively during sleep, when the force to the brain is turned off and consciousness ceases.*

Dreaming is the consequence of the partial supply of force to the brain, some parts of it being thus brought into activity while others are asleep. The greatest incongruity in thought and feeling is the consequence, and yet the world

* Mr. Charles Moore, in an Essay "On Going to Sleep," brings together, from different quarters, various facts which appear to him to justify the conclusion that a contraction of the cerebral arteries, shutting off to some extent the supply of blood to the brain, is the physical occasion of sleep; this contraction being due to the unimpeded energy of the sympathetic ganglia which comes into play when the inhibitory action of the brain upon them is withdrawn. He conjectures, also, that states of somnambulism and double consciousness may be produced by the separate contraction of particular arteries, the area of the brain dependent upon other arteries not being asleep.

of our dreams is quite as real to us as as that of our waking hours. Sometimes, the force being absent from other parts of the brain, is supplied in unusual quantities to the intellectual faculties, and there is a greater intellectual vigour than when awake. Thus one-third of our time we are liable to a derangement of the mental powers, which would place us in a lunatic asylum if it occurred during our waking hours. One-third of our existence we are either mad or dead, *i.e.*, dreaming or insensible.

The world, as we have already shown, is created within us. Certain forces derived from the food set the brain in motion, and other forces from without, coming through the senses, act upon these, and produce within us perception, or ideas, or nervous impressions of colour, form, size, a sense of resistance or weight, order, number, relative position, motion, likeness and unlikeness, and connection. These are called ideas of Simple and Relative perception, and Reflective Faculties. Simple ideas are real entities, formed or conditioned by the brain, and having no resemblance to anything out of the mind, or in the world: these simple ideas are by other faculties of the mind worked up into our conception of the world: for what would the world be without colour, or form, or size, etc.? and yet colour, form, size, etc., are ideas or feelings, not things.*

* The world objectively is Force and its correlations, or transmutations and modes of action; and Miss Martineau says: "Nothing in the experience of my life can at all compare with that of seeing the melting away of the forms, aspects, and arrangements under which we ordinarily view nature, and its fusion into the system of forces which is presented to the intellect in the magnetic (mesmeric) state."—("Letters on Man's Nature and Development," p. 122.) Now Force is the force of Mind—that is, Mind is Force; and Mr. Atkinson says, "when etherised myself, I felt all nature dissolved away, leaving only Mind."—(Ibid, p. 152.) And yet both Miss Martineau and Mr. Atkinson, in their natural state, are still what are called Materialists.

Properties, which we call of matter, are separate and distinct forces, but they are united by association in the mind, and rarely act singly, one generally calling the other into activity. Unity is also given to them in the mind, that is, by the faculty of Individuality, and we say these properties belong to body or matter. The same faculty individualises or gives unity to our separate ideas and feelings, and we say they belong to the mind; and more, we individualise each separate act of power or ability, dress this image in our attributes, and thus create a god after our own likeness, with our passions and modes of thought.

It is this unity of body and mind, which is a mere form of thought, to which we give the name of "I," and which assumes importance in proportion as it is associated with more or less of the feeling of self-esteem. All that is meant by this "Ego" is the succession of ideas and feelings which constitutes consciousness. "I am," means only, these ideas and feelings are.* "I think," is a pure assumption, or

* Huxley says, "Some of these states we refer to a cause we call 'self;' others to a cause or causes which may be comprehended under the title of 'not-self.' But neither of the existence of 'self,' nor that of 'not-self,' have we, or can we by any possibility have, any such unquestionable and immediate certainty as we have of the states of consciousness which we consider to be their effects. They are not immediately observed facts, but results of the application of the law of causation to those facts. Strictly speaking, the existence of a 'self' and of a 'not-self' are hypotheses by which we account for the facts of consciousness. They stand upon the same footing as the belief in the general trustworthiness of memory, and in the general constancy of the order of nature—as hypothetical assumptions which cannot be proved, or known with that highest degree of certainty which is given by immediate consciousness; but which, nevertheless, are of the highest practical value, inasmuch as the conclusions logically drawn from them are always verified by experience.

"This, in my judgment, is the ultimate issue of Descartes' argument; but it is proper for me to point out that we have left Descartes himself some way behind us. He stopped at the famous formula, 'I think, therefore I am.' But a little consideration will show this for-taking for granted of the "I;" all we know is, thinking is. The percipient, as when we say "I perceive," is one class of facul-

mula to be full of snares and verbal entanglements. In the first place, the 'therefore' has no business there. The 'I am' is assumed in the 'I think,' which is simply another way of saying 'I am thinking.' And, in the second place, 'I think' is not one simple proposition, but three distinct assertions rolled into one. The first of these is, 'something called I exists;' the second is, 'something called thought exists;' and the third is, 'the thought is the result of the action of the I.'

"Now, it will be obvious to you, that the only one of these three propositions which will stand the Cartesian test of certainty is the second. It cannot be doubted, for the very doubt is an existent thought. But the first and third, whether true or not, may be doubted, and have been doubted. For the assertor may be asked, How do you know that thought is not self-existent; or that a given thought is not the effect of its antecedent thought or of some external power? and a diversity of other questions, much more easily put than answered. Descartes, determined as he was to strip off all the garments which the intellect weaves for itself, forgot this gossamer shirt of the 'self;' to the great detriment, and indeed ruin, of his toilet when he began to clothe himself again."

Prof. Hegel holds: First, that thinking thinks; that there is no thinker—no Ego, except what we call thought; that Thinking and Thinker are but the two names for the same thing—not that that which is thought of is also that which thinks of it. Second, that every object of thought is a phenomenon, that the subject and the object are inseparable, that to be thought of (or to be perceived) and to be, are correlative terms; that as there can be no perceiving without an immediate object, so there can be no immediate object—no sound, no colour, for instance, without there being some act of perceiving—*i.e.*, of thinking, associated with it.—("Hegel and his Connection with British Thought," by Dr. T. Collyns Simon. *Contemporary Review*, January, 1870.) Hegel held that there is nothing but thought, and that thought and force are functions of the Unknowable. Mr. Lewes remarks on Hegel, in his "Biographical History of Philosophy:"—"But, the system itself we may leave to all readers to decide whether it be worthy of any attention, except as an illustration of the devious error of illustration. A system which begins with assuming that being and non-being are the same, because being in the abstract is the unconditioned, and so also is non-being; therefore both, as unconditioned, are the same—a system which proceeds upon

ties acting upon the others—it is reflection on consciousness, *i.e.*, on perception: one part of the mind becomes an "object"

the identity of contraries as the method of philosophy; a system in which thought is the same as the thing, and the thing is the same as the thought; a system in which the only real positive existence is that of simple relation, the two terms of which are mind and matter—this system, were it wholly true, leaves all the questions for which science is useful as a light, just as much in the dark as ever, and is, therefore, unworthy the attention of earnest men working for the benefit of mankind."

It is impossible to say what the unconditioned is *in itself*, but Mr. Lewes must admit that it is *nothing* to us, and, therefore, the same as "non-being." But we cannot accept Mr. Lewes's estimate of the worth of the Hegelian Philosophy. A system in which "thought is the same as the thing, and the thing is the same as the thought," is simply putting physics and metaphysics on the same inductive foundation, and showing that the assumed difference between them is not founded in fact. "This absolute and unholy barrier set up between Psychical and Physical nature must be broken down." Physics is the study of forces and their mode of action; metaphysics is the study of the same forces and their mode of action when made conscious by passing through the brain. Force, or the unknown entity of which it is the force, is thought, and thought is force—the thought is the same as the thing, and the thing is the same as the thought, *i.e.*, mind, conscious and automatic.

According to Schelling, "Nature is Spirit visible; Spirit is invisible Nature: this absolute Ideal is at the same time the absolute Real.

"Hence Philosophy has two primary problems to solve. In the *Transcendental Philosophy* the problem is to construct Nature from Intelligence—the Object from the Subject. In the Philosophy of Nature the problem is to construct Intelligence from Nature—the Subject from the Object. And how are we to construct one from the other? Fichte has taught us to do so by the principle of the identity of Subject and Object, whereby the productivity and the product are in constant opposition, yet always one. The productivity is the activity in act; it is the force which developes itself into all things. The product is the activity arrested and solidified into a fact; but it is always ready to pass again into activity. And thus the world is but a balancing of contending powers within the sphere of the absolute."—("Biographical History of Philosophy," p. 594.) This is exactly what we have shown it to be. Again: "Psychologists tell

to itself. It is this also that acounts for the unity of consciousness; for although our ideas and feelings are simultaneously various, the reflective faculty that attends to them is one, and they can therefore only be attended to in succession one at a time, whether they be simple or complex.

But it is said that "how consciousness arises cannot be explained, either from the scientific or from the philosophical side." But is this so? I think the knowledge we now have admits of its explanation—at least so far as we can explain anything. Force, or automatic mind, passing through the brain, by means of the peculiar molecular action of each nervous centre, resumes its conscious as an idea or feeling, and passes from unconscious to consciousness, and back again, as illustrated by the action of our own minds. A certain amount of vital energy or force passing through Destructiveness gives rise to the feeling of anger, and this concentrated energy is all accounted for when it returns through the muscles and knocks a man down in consequence. The brain is acted upon through the eye or the ear, pulls the trigger called the Will, and releases a reservoir of internal force, derived from the food, from the region of the brain between the ears, which passing along the motor-nerves of the arm, expends

us that there are three things implied in one act of vision, viz., a tree, an image of that tree, and a mind that apprehends that image. Fichte tells me that it is I alone who exist: the tree and the image of the tree are but one thing, and that is a modification of my mind. This is *Subjective Idealism*. Schelling tells us that both the tree and my Ego are existences equally real or ideal, but they are nothing less than manifestations of the Absolute. This is *Objective Idealism*." And this is our idea. "But, according to Hegel, all these explanations are false. The only thing really existing (in this one fact of vision) is the Idea—the relation. The Ego and the Tree are but two terms of the relation, and owe their reality to it. This is Absolute Idealism."—("Biographical History of Philosophy," p. 607.) The force reflected from the object or tree, meeting the force within the brain, or the Ego, passes into consciousness, forming the idea. This is the fact of which Hegel had at least a dim perception.

itself upon the cause of anger. Whether we shall ever, as Prof. Huxley supposes, arrive at an exact mechanical equivalent for facts of consciousness, it is impossible to say; although it is highly probable we are far from it at present. Huxley says: "I believe that we shall, sooner or later, arrive at a mechanical equivalent for facts of consciousness, just as we have arrived at a mechanical equivalent of heat. If a pound weight, falling through a distance of a foot, gives rise to a definite amount of heat—which may properly be said to be its equivalent—the same pound weight, falling through a foot on a man's hand, gives rise to a definite amount of feeling—which might, with equal propriety, be said to be its equivalent in consciousness." It will probably be later rather than sooner that this problem will be solved, for it is much complicated by the fact that the consciousness evolved is the joint produce of the brain or nervous force, and that from the pound weight, the one derived from the arrested motion of the atoms of oxygen and carbon, as they clash together in the body; the other from the arrested motion of the pound weight.

"The broad distinction between nerve-energy and mind-energy has been distinctly maintained by many of the most distinguished promoters of scientific inquiry." *

The sequence and variation of our consciousness, which we call modes of action of the mind, are divided into Perception, Conception, Memory, Imagination, and Judgment. The first of these—perception, is composed of the force within and the force from without, and therefore has a double intensity to any of the others, which proceed from the action of the brain alone without the sense. Intensity of thought and feeling is always in proportion to the amount of force expended in producing it, and this furnishes a good criterion of externality, that is, of the difference between the ideas we receive from

* *Contemporary Review*, "Physical Science and Mental Philosophy," p. 230, January, 1871.

without through the senses and those that are the production of the mind alone. The external world has thus double the reality of mere ideas: a child lives a life of sensation and perception, an old man one of ideas. The difference between Primary and Secondary qualities is not that generally recognised by metaphysicians between colour and extension, &c., as the latter is as much in the mind as the former, but between the ideas of simple and relative perception—that is, between the ideas received directly from without and those manufactured in the mind out of these.

Belief—does it belong to the intellect, the emotions, or the will? James Mill resolves it into purely intellectual elements, but it is a feeling—as Hume says, an act of the sensitive part of our nature, rather than the cogitative, and we necessarily believe in the Intuitions of our faculties. We take the results of their action for granted as fundamental truths, and as nature obliges us to eat without reasoning upon its necessity, so she obliges us to believe what is equally necessary for our action and well-being, without reference to its absolute truth or even speculative reasonings. We reason and doubt afterwards, and begin to suspect that our faculties do not always tell us the truth. We believe in the real existence of the external world of our consciousness, and do not doubt till we find it is created within us. We believe in matter and in mind, and not in a correlation of forces or manifestation of Divine Power. We believe that every effect has a cause, but that is no reason why it should be true; there may be only sequence. We believe in Space as an objective reality, when, like motion, it is nothing in itself—it is that in which an entity exists, and is as inseparable from the entity as motion from the thing moving, or power from that of which it is the power.

As to the *necessary* connection between cause and effect, of unalterable laws, of the immutability of the laws of nature, consciousness tells us nothing about this. The higher probability is that this connection has been established for a purpose, and will continue as long as that purpose is required. J. S. Mill says: "Any *must* in this case, any necessity, other than the unconditional universality of the fact, we know nothing of. * * All I know is, that it always does." Reason is based upon the invariableness, upon the knowledge that what has taken place, in the same circumstances and under the same conditions, will take place again. Law, in my opinion, as I have said before, is caused by automatic Will-power, and is changeable should circumstances or conditions require. Oersted says, "the world is governed by an eternal reason, which makes known to us its actions by unalterable laws." I object to such terms as "eternal" and "unalterable," as we can know nothing about them. As far as we know, the same laws exist throughout the universe, but that is no reason why they should continue to do so. Laws of nature are attributes of Deity, and the manifestations, if not the attributes, are ever changing.

The sense of Personal Identity is, like faith or belief, a feeling, and not a mere intellectual perception. It is an instinct, an intuition, a pure creation of the mind. The "I" of which we are conscious, and which we say "thinks, wills, and feels," is, as we have seen, the aggregate of our bodily and mental states, to which unity has been given by a faculty of the mind. Any kind of sameness or identity must be a delusion, an anthopomorphism, for we are never the same for any two seconds together. In both mind and body we are part and parcel only of the ever-varying forces around us. This notion of "self" and identity is not dependent upon memory, for it remains when all memory is lost; past sensations give no such feeling. That it is dependent upon a condition of the brain is evident, as under certain states of excitement this sense is lost altogether, as in the case of the two gentlemen who had been dining out: in passing through a ford one of them fell into the water and called out to his companion, "I say, Bill, there is *some one* fallen into the water." The feeling also of personality is sometimes "double," and otherwise deranged.

Having seen, then, Whence comes Consciousness, we have now to inquire, What is the object of it? This brings us face to face with final cause, and the question is, Do we know anything, or can we know anything of final cause? The *Athenæum*, in its review of Prof. Owen (July 17, 1869,) says, "long ago philosophers have proved both final causes, and innate ideas and tendencies, to be mere assumptions." Dr. Büchner says, "the motion of matter is as eternal as matter itself. Why matter assumed a definite motion at a definite time is as yet unknown to us;" but if matter ever *began* to move towards any definite aim, this is admitting voluntary action and intelligence, which I think the Doctor denies. The question then is, what was the object when matter began to move towards the production of organised beings? Is it, as the Doctor says, unknown to us? We have already considered this subject, and, as I have said, a world without consciousness is practically non-existent—the same as no world at all; and that therefore the production of consciousness is more desirable than the production of a world. But consciousness that was merely cognisant and took no interest would be no better than none at all; it must be either pleasurable or painful. No one desires a painful consciousness, and a pleasurable consciousness therefore is the only desirable end in the production of sensitive organised creatures. This is the final cause; this is "why matter assumed a definite motion at a definite time," and things are right or wrong as they tend towards this object—that is, towards pleasurable consciousness.

The world, as we conceive of it, is no more a reality than our dreams; each creature creates its own world and carries it about in its own head; outside itself there is nothing but the play of forces on the nervous centres of each being. This creates a thousand worlds adapted to the different wants of each, and is so much better, therefore, than if it had objective reality; and man stupidly thinks it was all made for him, and that the way these forces affect him is the absolute measure of truth: whereas a very few only of the forces around him reach him through his five senses or otherwise, and produce those impressions and ideas which are sufficient to guide him towards the objects of his being, towards his *real* world—that of his pleasures and pains, and which he calls his moral world. In reality he knows very little indeed of all that is going on without him, as he is cognisant only of the influences that can penetrate through his thick skin; and whole worlds of beings may exist *without* his intellectual ken. If our faculties, then, are few and limited, and not designed to penetrate into the inward essence and constitution of things, they are yet sufficient for our purpose, the use of our intellectual consciousness being not to teach absolute truth, but to guide us towards our wants, which in their fulfilment are always pleasurable, and thus contribute towards the stock of happiness in the world.

Good and evil are purely subjective, and the moral world is as entire a creation of the mind as the physical world. It is merely a record of man's pleasures and pains, of his likes and antipathies, and of the various fine names by which he distinguishes the different varieties of feeling as he wishes to promote the one and to prevent the other. As our thoughts and ideas compel a reference to objects out of self, so do our feelings, and we talk of the eternal and immutable distinctions between right and wrong, whereas these distinctions have no existence out of ourselves, and one action is as good as another *in itself*, and is good, pure, holy, &c., in proportion as it tends to carry out the purpose of creation, which is not man's happiness alone, but that of all of sensitive existence. Morality is the science of living together in the most happy manner possible; at present it is confined to men alone, but we must widen its sphere of action so as ultimately to take in all living creatures. Do not let us be be alarmed, then, for the interests of morality, for as J. S. Mill says, "a volition is a moral effect, which follows the corresponding moral causes as certainly and invariably as physical effects follow their physical causes."

Physical science has made rapid progress since the introduction of the inductive method, while mental science, to which it is supposed not to apply, is little further advanced than it was two thousand years ago; but on the recognition of this great truth, that causation is as constant, and that law reigns as much in the realm of mind as of matter, our future progress in this department must depend. This truth occupies, in the present day, much the same position in mental science, as the earth's position with respect to the sun in the days of Copernicus did in physics. Men saw that the sun went round the earth, and the Bible said it did, and Galileo was imprisoned for saying it did not. Men say they now *feel* that they are free, as they before *saw* that the sun went round the earth; and theologians say that responsibility, which, according to them, is the right to take revenge for past misconduct, depends upon this freedom, and that morality depends upon this kind of responsibility; and when our philosophers are appealed to as to whether man is free, or subject to law, like everything else, they say, "Sometimes one, and sometimes the other." To give an instance in each class: Froude, the philosopher, says: "The foolish and the ignorant are led astray by the idea of contingency, and expect to escape the just issue of their actions; the wise man will know that each action brings with it its inevitable consequences, which even God cannot change without ceasing to be himself." * Praise and blame "involve that somewhere or other the influence of causes ceases to operate, and that

* Froude's Essays, "Spinoza," vol. ii, p. 48.

some degree of power there is in man of self-determination, by the amount of which, and not by their specific actions, moral merit or demerit is to be measured." * How "inevitable consequences" are to be expected where "causes cease to operate," he does not tell us; but no doubt the earth goes round the sun in physics, and the sun round the earth where man's volition is concerned. Huxley, the man of science, says: "Theology in her purer forms, has ceased to be anthropomorphic; however she may talk. Anthropomorphism has taken her stand in its last fortress—man himself. But science closely invests the walls; and philosophers gird themselves for battle upon the last and greatest of all speculative problems. Does human nature possess any free volitional or truly anthropomorphic element, or is it only the cunningest of all nature's clocks? Some, among whom I count myself, think that the battle will for ever remain a drawn one, and that, for all practical purposes, this result is as good as anthropomorphism winning the day." †

Notwithstanding, we are slowly, but surely, coming to the conviction that in nature there is no beginning,—merely pre-existent and persistent force and its correlates—that is, "that each manifestation of force can be interpreted only as the effect of some antecedent force, no matter whether it be an inorganic action, an animal movement, a thought or feeling"; ‡ that all force, or power, or ability is derived and inseparable from that of which it is the force—the Supreme Cause of all. If we have lost matter, we have found force; if we have lost mind—a supposititious, capricious existence, *governed* by nothing—we have found universal law, and "a supreme and infinite and everlasting Mind in synthesis with all things." In the correlation of force, we have one great heart-beat of the Absolute Existence. "Being underlies all

* Froude's Essays, "Spinoza," vol. ii., p. 59.

† *Fortnightly Review*, June, p. 664.

‡ "First Principles," by Herbert Spencer.

modes and forms of being."* "Nature is an infinitely divided God. * * The Divine One has dispersed itself into innumerable sensible substances, as a white beam of light is decomposed by the prism with seven coloured rays. And a divine being would be evolved from the union of all these substances, as the seven coloured rays dissolve again into the clear-light beam. The existing form of nature is the optic glass, and all the activities of spirit are only an infinite colour-play of that simple divine ray."†

* Hegel. † Schiller.

## CHAPTER V.

### RELIGION.

We have seen how the physical and moral worlds have been created within us by our forms of thought, and by out likes and antipathies; the latter constituting a World based on the Chemistry of Sensibility. We have now to consider in what way and by what faculties the Religious World is created. Morality defines our relation to our fellow-man, religion our relation to God. The properties of matter, regarded as separate and distinct forces acting upon our organisation, give rise to our "forms of thought." The mental faculty of Individuality gives unity to these properties of matter, creating what we call substance; the same faculty gives unity to our separate thoughts and feelings, and to that unity we give the name of Mind; but the process does not stop here; for every change or effect we see that there is a cause which is always equal, under like circumstances, to produce the same effect. This force, or this power or ability to produce change, is unseen and unknown to us except in its effect, but we necessarily conclude that it is the force of something, and Individuality again gives unity to all these separate powers or causes of effects, and to this unity we give the name of God. Examination into the nature of this Force shows that it is Persistent,—that is, is the effect, or correlate, or equivalent of some antecedent force, and that what we call matter and mind are known to us only from the changes that take place in it, its mode of action or phenomena: "in other words, matter and spirit are but names for the imaginary substrata of groups of natural phenomena."* These changes or phenomena do not persist, but

* Huxley.

pass away, but their Unknown Cause, or the power that produced them, continues to exist, and as far as we know has existed, and will continue to exist, without beginning or end. This Unknown Cause—this abstract conception of Force or Power—we immediately proceed to invest with all our own modes of thought and feeling, and as these have varied with the progress of intelligence, so have the attributes we have ascribed to the unknown Source.

Although "the Reality behind the veil of appearances" is unknown to us, yet we infer its nature from its manifestations, and its character, as we judge of all character, by the tendency of its actions.

This idea of Supreme Power, originating in the intellect, is the base of all Religion; although the Feelings to which it gives rise and with which it is associated more properly constitute Religion. Thus, as hidden power and as the cause of all things it excites Wonder; as supreme power it creates Veneration or Reverence, Hope, and Fear. Again, we have a strong "love of life," and a great dread of ceasing to be; we cannot bear the thought of parting *for ever* from those who are dear to us; we have an intuition of persistent force—of the something that continues to exist independently of all change in ourselves, which we feel is indestructible, and which we call our immortal soul; and as we strongly wish for another world, or state of existence after it is apparent we have done with this, our feelings of Hope and Faith give reality to what we so much desire, and we believe in this other world. Each religion, then, peoples this future state with beings of its own creation. Strong feelings beget myths. The Bard and the Prophet in the earliest ages were the same; the Bard expressing and harmonising public feeling, and the people necessarily believing as inspired that which so harmonised. In this way religions are formed and religious feeling is generated.

It is said that this dread of ceasing to be, and this yearning for a "future state," are evidence that that future state is to be, since God would not have implanted within us a feeling that was not to be gratified, and "certain it is that man alone yearns for something that neither sense nor reason can supply;" but we must recollect that such feelings have their uses in this life, and that this dread of ceasing to be induces us to do all we can to keep both ourselves and our friends in existence here, and that therefore the transference of such feelings to another world may be a misdirection of them. Veneration also, which is worship as applied to God, has also its useful application here, as it is the source of politeness and of that courtesy which we owe and pay to one another. Society could scarcely exist without it; in fact, what are called the Religious Feelings—Veneration, Hope, Wonder or Faith—have, as we have previously seen, their application here.

This hidden, and therefore mysterious, Power—a Power behind and beyond and exceeding his own in all his chief interests, appears to have been recognised in all stages of man's existence. At first there were supposed to be separate powers for all the more mysterious and powerful phenomena in nature, and we had gods of the earth and of the heavens, of the winds and of the waters, of war and of wisdom; and unity was only given to these powers with increasing intelligence; * and thus unity can only be said to be recently proved

* In the Rigveda, a collection of ancient Hindu Hymns, &c., the old authority for religious and social Institutions, made 2,000 years before Christ, the principal deities mentioned are Agni, the god of fire; Indra, the regent of the firmament, who shatters the clouds with his thunderbolt, and dispenses rain; Nitra and Varuna, the representatives of day and night; Sûrya, the sun; Vayu, the wind; the Maruts, or tempests; Asvins, &c., together with Vishnu and Rudra, who play so prominent a part in the later mythology, but so small in the Vedic pantheon.

in the modern discovery of Pre-existent and Persistent Force and its Correlates.

It is said that God made man in his own image. It is certain that man has made God in *his* own image: the character and attributes thus given to him varying from those of the lowest savage to the highest ideal of the present day.* Fear of this unknown power at first predominated over hope and love. The source of much that was good and evil was hidden, and people feared the evil,† the attention being most called to it; good being the rule, evil the exception. Everything was done to propitiate the supposed source of this evil:

* "The 'stages' of religious thought, according to Sir John Lubbock are:—*Atheism*, understanding by this term not a denial of the existence of deity, but an absence of any definite ideas on the subject; *Fetichism*, the stage in which man supposes he can force the deity to comply with his desires; *Nature Worship*, or Totemism, in which natural objects—trees, lakes, stones, animals, &c.—are worshipped; *Shamanism*, in which the superior deities are far more powerful than men, and of different nature, their place of abode is also far away, and accessible only to Shamans; *Idolatry*, or Anthropomorphism, in which the gods took still more completely the form of men, being, however, more powerful. They are still amenable to persecution; they are a part of nature, and not creators; they are represented by images or idols. The next stage the deity is regarded as the author, not merely a part, of nature; he becomes for the first time a really supernatural being. The last stage to which I will refer is that in which morality is associated with religion."—(*The Student*, October, 1870, p. 398.)

† "If the mere sensation of fear, and the recognition that there are probably other beings more powerful than himself, are sufficient alone to constitute a religion, then we must, I think, admit that religion is general to the human race. But when a child dreads the darkness and shrinks from a lightless room, we never regard that as an evidence of religion. Moreover, if this definition be adopted, we can no longer regard religion as peculiar to man. * * * The baying of a dog to the moon is as much an act of worship as some ceremonies which have been described as religion by travellers."—("The Origin of Christianity," by Sir John Lubbock, Bart., pp. 124-5.)

the best of all the flocks and herds, and even human beings, were offered to it; and particularly, as praise and worship were most acceptable to themselves, men thought they could not be less acceptable to the mysterious Power they were anxious to propitiate. They flattered their gods, as men do one another, for what they could get out of them. They called them all sorts of fine names, thinking it would please them. They were thought to be immoderately fond of power and glory, like some poor vain mortal, and they were even supposed to have created the world, not for the good of the creatures on it, but for their "glory," as a Frenchman might have done. The reflex of such childish notions is to be found even in the present day. Sir John Lubbock tells us (British Association Meeting, 1870,) that "the natives of Kamschatka insult their deities if their wishes are unfulfilled: they even feel a contempt for them. If Kutka, they say, had not been stupid, would he have made inaccessible mountains and too rapid rivers? The Lapps made images of their gods, putting each in a separate box, on which was written the name of the deity, so that each might know his own box."* This

* At the close of his lecture on "Savages" Sir John Lubbock said:— "Gradually, however, an increased acquaintance with the laws of nature enlarged the mind of man. He first supposed that the Deity fashioned the earth, raising it out of the water, and preparing it as a dwelling place for man; and subsequently realised the idea that land and water were alike created by Divine power. After regarding spirits as altogether evil, he rose to a belief in good as well as in evil deities, and, gradually subordinating the latter to the former, worshipped the good spirits alone as gods, the evil sinking to the level of demons. From believing only in ghosts, he came gradually to the recognition of the soul: at length uniting this belief with that in a beneficent and just Being, he connected morality with religion, a step the importance of which it is scarcely possible to over-estimate. Thus we see that as men rise in civilisation their religion rises with them; that far from being antagonistic to religion, without science true religion is impossible.

also is exactly what is going on among more civilised people: each sect makes a temple for its God, hoping that he will know his own, and that the mistakes that have, in some unaccountable way, been made in this world, will be rectified in another. God's government is even supposed not to begin in this world, but in the next; but if, as theologians affirm, inherent disorder reigns here, what just and sufficient reason is there to suppose that order reigns elsewhere?

The Jews are supposed to have had the most perfect conception of this Supreme Power, but the Hebrew Scriptures represent their God as existing in the form of a man; walking in the Garden of Eden in the *cool* of the evening; and on one occasion it is related of Moses that he saw a portion of God's person like the hinder parts of a man. But not only do the Jews suppose themselves to be made bodily in the likeness of God, but they give to God the likeness of all their worst passions. He is unjust, visiting on one man the sins of another. He is proud, jealous, revengeful, the commander-in-chief of their armies, ordering whole towns and peoples to be put to death—men, women, children, and even the cattle—because they called *their* god by another name; and ordering the sun and moon to stand still that the Jews might make a greater slaughter of their enemies; but at the same time easily turned from His purpose by obedience and flattery. To these vices were certainly added all the human virtues, and some of the highest attributes which the human mind is capable of conceiving, and several that are beyond man's conception.

The Christian God is all this, and, if we are to judge from the prevalent creeds, *infinitely* worse. The ancient Hebrews, at least, believed that evil was confined to this world, not believing in a future state; but the Christians make evil Absolute, as, according to them, the torture of the wicked and unbelieving, who are supposed to be the great majority, is to endure *for ever and ever*. The Creed of St. Athanasius, read in all our churches, says:—

"Whosoever will be saved; *before all things* it is necessary that he hold the Catholic Faith.

"Which Faith, except every one do keep whole and undefiled: *without doubt* he shall perish everlastingly. * * *

"He (Christ) suffered for our salvation: *descended into hell*, rose again the third day from the dead.

"At whose coming all men shall rise again *with their bodies:* and shall give account for their own works.

"And they that have done good, shall go into life everlasting: and they that have done evil, into everlasting fire.

"*This* is the Catholic Faith, which except a man believe faithfully he cannot be saved.

"Glory be to the Father, and to the Son: and to the Holy Ghost: As it was in the beginning, is now, and ever shall be: world without end. *Amen*."

This ready accession, this amen, or so be it, in so awful a sentence—this thankfulness even, or "glory to God," would be ludicrous, if it were not really lamentable, as indicative of the greatest possible debasement and servility of mind. Belief is not voluntary: we cannot believe black to be white, whatever may be the consequences. And it is precisely the same in all minor degrees of faith: we can only believe that which is credible, and which approves itself to us as truth, whether we "perish everlastingly" or not.

But it may be supposed that this is a Creed that does not approve itself to Christians of the present day. Quite the contrary. The Church is by no means prepared to give up this cheering article of faith. In the debates in the House of Lords (March, 1871,) on the Lectionary Bill, in which bill the Athanasian Creed was retained, Earl Grey characterised this Creed "as the barbarous production of a barbarous age, to which he never listened without feelings of horror and disgust." Earl Beauchamp, on the contrary, described it as a source of comfort and consolation to many generations of churchmen. Upon which the *Pall Mall Gazette* remarked that "Those to whom it is a 'source of comfort and consolation' to anathematise large portions of the human race may surely fortify themselves in this way in private without compelling the participation of those who find in the repetition of comprehensive anathemas nothing which either comforts or consoles." The Church is even less prepared to give up the *eternity* of punishment. When some time since, on the publication of "Essays and Reviews," the Church was thought to be in danger, ten thousand of the clergy "held it to be their bounden duty to the Church, and to the souls of men," to declare their firm belief that the "punishment" of the "cursed," equally with the "life" of the righteous, is "everlasting." Max Müller says: "There is no religion—or if there is I do not know it—which does not say, 'Do good, avoid evil.' There is none which does not contain what Rabbi Hillel called the quintessence of all religions, the simple warning, 'Be good, my boy.' 'Be good, my boy,' may seem a very short catechism; but let us add to it, 'Be good, my boy, for God's sake,' and you have in it very nearly the whole of the Law and the Prophets."* The principal article of faith, however, of the English religion is, "Be good, my boy, *or Bogie 'll have you.*" The magistrate asks, as the sole test of orthodoxy, "Do you believe in a future state of rewards and punishments? Do you know where you will go to if you do not tell the truth?"

The Roman Catholic holds the same doctrine, and Mr. Furness, "*permissu superiorum*," informs us that "as the devil is king of hell, he is also judge. When a soul comes into hell, condemned by the judgment of God, he executes the judgment. He fixes whereabouts in hell the soul is to be, how it is to be tormented, and what devils are to torment it."

* "Science of Religion," *Fraser*, July, p. 104.

Of the Scotsman's idea of God we may judge from the Westminster Confession of Faith: this tells us that

"God from all eternity did, by the most wise and holy counsel of his own will, freely and unchangeably ordain whatsoever comes to pass: yet so, as thereby neither is God the author of sin, nor is violence offered to the will of the creatures, nor is the liberty or contingency of second causes taken away, but rather established.

"Although God knows whatsoever may or can come to pass upon all supposed conditions; yet hath he not decreed any thing because he foresaw it as future, or as that which would come to pass upon such conditions.

"By the decree of God, for the manifestation of his glory, some men and angels are predestinated unto everlasting life, and others foreordained to everlasting death.

"These angels and men, thus predestinated and foreordained, are particularly and unchangeably designed; and their number is so certain and definite, that it cannot be either increased or diminished.

"Those of mankind that are predestinated unto life, God, before the foundation of the world was laid, according to his eternal and immutable purpose, and the secret counsel and good pleasure of his will, hath chosen in Christ unto everlasting glory, out of his mere free grace and love, without any foresight of faith or good works, or perseverance in either of them, or any other thing in the creature, as conditions, or causes moving him thereunto; and all to the praise of his glorious grace.

"As God hath appointed the elect unto glory, so hath he, by the eternal and most free purpose of his will, foreordained all the means thereunto. Wherefore they who are elected being fallen in Adam, are redeemed by Christ; are effectually called unto faith in Christ by his Spirit working in due season; are justified, adopted, sanctified, and kept by his power through faith unto salvation. Neither are any other redeemed by Christ, effectually called, justified, adopted, sanctified, and saved, but the elect only.

"The rest of mankind, God was pleased, according to the unsearchable counsel of his own will, whereby he extendeth or withholdeth mercy as he pleaseth, for the glory of his sovereign power over his creatures, to pass by, and to ordain them to dishonour and wrath for their sin, to the praise of his glorious justice."

And of the state of men after Death, and of the Resurrection of the Dead, the same Confession of Faith tells us

"The bodies of men after death return to dust, and see corruption; but their souls, (which neither die nor sleep,) having an immortal subsistence, immediately return to God who gave them. The souls of the righteous, being then made perfect in holiness, are received into the highest heavens, where they behold the face of God in light and glory, waiting for the full redemption of their bodies; and the souls of the wicked are cast into hell, where they remain in torments and utter darkness, reserved to the judgment of the great day. Besides these two places for souls separated from their bodies, the scripture acknowledgeth none.

"At the last day, such as are found alive shall not die, but be changed: and all the dead shall be raised up with the self-same bodies, and none other, although with different qualities, which shall be united again to their souls for ever.

"The bodies of the unjust shall, by the power of Christ, be raised to dishonour; the bodies of the just, by his Spirit, unto honour, and be made conformable to his own glorious body."

And of the Last Judgment we are told

"God hath appointed a day wherein he will judge the world in righteousness by Jesus Christ, to whom all power and judgment is given of the Father. In which day, not only the apostate angels shall be judged, but likewise all persons that have lived upon earth shall appear before the tribunal of Christ, to give an account of their thoughts, words, and deeds, and to receive according to what they have done in the body, whether good or evil.

"The end of God's appointing this day is for the manifestation of the glory of his mercy in the eternal salvation of the elect, and of his justice in the damnation of the reprobate, who are wicked and disobedient. For then shall the righteous go into everlasting life, and receive that fulness of joy and refreshing which shall come from the presence of the Lord; but the wicked, who know not God, and obey not the gospel of Jesus Christ, shall be cast into eternal torments, and be punished with everlasting destruction from the presence of the Lord, and from the glory of his power."

Of the Fall of Man this confession says:

"Our first parents being seduced by the subtilty and temptation of Satan, sinned in eating the forbidden fruit. This their sin *God was* *pleased, according to his wise and holy counsel, to permit, having purposed to order it to his own glory!!"**

To show that these creeds still belong to the present time, take the following illustration:—The Rev. Fergus Ferguson, of Dalkeith, had been indulging in certain interpretations of Scripture and the Westminster Confession, which, in May, 1871, brought him under the lash of the United Presbyterian Synod, the governing body of what is called in Scotland the U. P. Church. The following propositions were submitted to Mr. Ferguson:—

"1. That all who shall ultimately be saved were chosen of God in Christ before the foundation of the world.

"2. That all who are saved are accepted of God, 'by His grace, through the redemption that is in Christ Jesus,' at or before the time of their death; and that none dying unsaved will, after death, have an opportunity of obtaining salvation.

"3 That whatever new revelations of Christ or of the truth are made after death to the saved are made, not to free them from sin, but to increase their knowledge and blessedness.

"4. *That, notwithstanding the inability of the will through sin, as taught in our Confession, unbelievers are fully answerable for their rejection of the offer of salvation which the gospel makes to them.*"

* Mr. Disraeli evidently takes a more hopeful view of things. The right hon. gentleman says: "If the Jews had not prevailed upon the Romans to crucify our Lord, what would have become of the Atonement? But the human mind cannot contemplate the idea that the most important deed of time could depend on human will. The immolators were preordained, like the victim, and the holy race supplied both. Could that be a crime which secured *for all mankind* eternal joy; which vanquished Satan and opened the gates of Paradise? Such a tenet would sully and impugn the doctrine that is the foundation of our faith and hope. Men must not presume to sit in judgment on such an act; they must bow their heads in awe and astonishment and trembling gratitude."—("Life of Lord George Bentinck," p. 489.) So that, according to the right hon. gentleman, Judas Iscariot and Pontius Pilate must have been the greatest benefactors the world has ever known, and "his holy race" supplied the former at least!

"These four propositions," says the report of the committee who framed them, "having been read over in Mr. Ferguson's hearing, he intimated his unqualified assent to them." And at the call of the committee, Dr. Cairns "offered thanks to God for the harmonious and happy result."

This is called the "Religion of Love," the "Gospel of Glad Tidings;" but may it not more properly be asked, Is it Theism or Demonism? We should have thought it was impossible to conceive anything so horrid—so vile a libel on our Creator. We are told that God from all eternity *ordained* the everlasting torment of creatures that he was at full liberty to bring into existence or not; and "according to this wise and holy counsel" allowed Satan, whom he also had created in the full knowledge of all the evil he would commit, to seduce our first parents, "having purposed to order it *to his own glory*." Also that the Almighty doomed some men from eternity to damnation, and then sent his Son on earth to mock them with the false promise of redemption he had previously decreed for them should never be; and this is called a great mystery. It is simply not true, and the only mystery about it is that any one, even in the most barbarous and savage age, could be found to believe it. "How," asks Froude, in his lecture on Calvinism (Mar. 17, 1871), "are we to call the Ruler who laid us under such an iron code, wise, or just, or merciful?" He is simply an almighty tyrant, who claims the right "to torture the innocent to all eternity." "There is one thing worse than a world without a sun—Humanity without a God," says the Rev. W. Jackson ("Positivism," p. 61); but how can humanity be without a God? A man by shutting his eyes does not produce "a world without a sun;" neither does a man who sees only what are called second causes obliterate a single ray of God's power and influence. One would think, from the language of theologians, that the very existence of God depended upon our *belief* in it, or that if we could not see God, He would be equally blind to us, when we know that "He makes His sun to shine on the evil and on the good." It is time this kind of thing was withdrawn from addresses to the enlightened men of the 19th century, and confined to the women. What we want above all things are truthfulness and sincerity, and there is no fear that frail Humanity will be without God, because of its blindness and shortsightedness. A child that cannot see its own father is to be pitied, and not left to grope about to its own destruction. Surely Atheism is infinitely preferable to this Calvinistic creed, and if we are to be damned for not believing it, all I can say is, I would rather be damned. "It were better," says Bacon, "to have no opinion of God at all, than such an opinion as is unworthy of him; for the one is unbelief, the other is contumely: and certainly superstition is the reproach of deity. Plutarch saith well to that purpose: 'Surely I had rather a great deal men should say there was no such man at all as Plutarch, than that they should say that there was one Plutarch that would eat his children as soon as they were born.'"* Canon Kingsley defines Superstition as "fear of the unknown," and he says "man can, what the lower animals (happily for them) cannot—organise his folly, erect his superstitions into a science, and create a whole mythology out of his blind fear of the unknown. And when he has done that—Woe to the weak!"† The fear of God originally meant literally "fear," not reverence, as it does now, and sacrifice and atonement were its offspring. Children and other human victims have been offered up by millions to this fear, and in the hope of propitiating this dread unknown; but it is impossible to conceive that it could have culminated in anything half so horrible as this Calvinistic creed. Its legitimate children were the Inquisition and Witchcraft.

* Essays, 17, "Of Superstition."

† "Superstition." A Lecture delivered at the Royal Institution, April 24, 1866. *Fraser*, June, 1866.

The Atheism of the present day is not so much a denial of the existence of the Universal Spirit of the Universe, as a protest against such views of God, and the too prevalent anthropomorphism. Prof. F. W. Newman says: "If a national religion be totally corrupt, widespread Atheism is nothing but the natural death of a creed which has lost moral vitality. If the Atheism springs from moral indignation, I believe that it can only be a temporary winter of the national soul in preparation for a more fruitful summer. If a very corrupt national creed—say, like that of Hindooism—were swept away by Atheism, when other agencies had failed, we perhaps ought to regard the Atheism as a beneficial visitation, like a hurricane which destroys pestilence."*

Theologians set up a baseless theory of their own, derived from a savage and barbarous age—an age of ignorance and superstition—and when facts do not correspond, it is regarded as a great mystery, which can never be solved, as we see only in part, and we are asked to put out our own eyes that we may see more clearly. "The idea of a universal and beneficent Creator of the universe does not seem to arise in the mind of man, until he has been elevated by long-continued culture; on the other hand, a belief in all-pervading spiritual agencies seems to be universal, but then they have generally been cruel and malignant spirits.†"

Lecky, speaking of the Christian ideas of Hell, says: "It is impossible to conceive more ghastly, grotesque, or material conceptions of the future world than they evince, or more hideous calumnies against that Being who was supposed to inflict upon his creatures such unspeakable misery;" and of the influence of such ideas, he says they were "well fitted to transform the man who at once realised it, and who accepted it with pleasure, into a monster of barbarity." According to the popular belief, all who differed from the

* "Causes of Atheism," Thomas Scott, Ramsgate.

† "Descent of Man," vol. 2, p. 395, C. Darwin.

teaching of the orthodox lived under the hatred of the Almighty, and were destined after death to an eternity of anguish, and no moral or intellectual excellence could atone for their crime in propagating error; and, according to *Saint* Gregory, the elect "will be sated with joy as they gaze on the unspeakable anguish of the impious, returning thanks for their own freedom. * * * The just man will rejoice when he shall see the vengeance."* Surely Nirwána were preferable to this. To this portion of the received Theology, viz., "that correct opinions are essential to salvation, and that Theological error necessarily involves guilt,"—may be traced, as Mr. Lecky says, "almost all the sufferings that Christian persecutors have caused, almost all the obstructions they have thrown in the path of human progress," religious ignorance and bigotry "adding immeasurably to the difficulties which every searcher after truth has to encounter, and diffusing far and wide intellectual timidity, disingenuousness, and hypocrisy."† The Shakers are the only people who can be said to have acted in any respect consistently with this creed. They had no children. "They had put to themselves this question: If all men born into the world are born into sin, and made the heirs of death in the world to come, how can the saint, when raised from his fallen nature, dare to augment this empire of sin and death?"‡ Here is the modern Christianity logically based on this creed:—

"Nothing is worth a thought beneath,
But how we may escape the death
That never, never dies.
How make *our own* salvation sure,
And when we fail on earth, secure
A mansion in the skies."

But the Rev. F. W. Robertson, of Brighton, says, "the popular religion represents only the female element in the national mind, at once devotional, slanderous, timid, gossip-

* Lecky's "Morals," vol., 2, p. 241. † Ibid. pp. 420, 376, 377.
‡ "New America," vol. 2, p. 116, by Hepworth Dickson.

ping, narrow, shrieking, and prudish."* The great Mr. Martin F. Tupper, that wonderful reflex of the popular mind, has also given us his "Creed." He solves the question of the origin of evil, and it is pleasant also to find that the poor ill-used brutes, as well as man, are to have *their* "future state" and full compensation. He says:

So then, in brief, my Creed is truly this,
Conscience is our chief seed of woe or bliss,—
God, who made all things, is to all things Love,
Balancing wrongs below by rights above,—
Evil seemed needful that the Good be shown,
And Good was swift that Evil to atone,
* * * * *
And the meek lamb may enter into life
Beside its butcher with his bloody knife;
That cruel costermonger and his drudge
Alike shall stand before their righteous Judge.

"No inquirer," says the Rev. James Martineau, "can fix a direct and clear-sighted gaze towards truth, who is casting side glances all the while on the prospects of his soul"; and Herbert Spencer as truly says it is impossible to believe in the "sinfulness of an unselfish sympathy or a pure love of rectitude—or that that conduct is only good whose only motive is other worldliness."†

The world has probably suffered as much from Christian Creeds and the dogmas of divinity as it has gained from the principles and precepts of the Christian Morality. It was held that "the Moors and Gentiles being condemned in the principal part, which is the soul, that part of them which it animates cannot be privileged with the benefit of our laws, because they are not members of the evangelical congregation."‡ In Spain dogma has spoilt one of the finest races

* "The Genius of Solitude," p. 324, by W. R. Alger.
† "First Principles," p. 120.
‡ Vasga de Gama's "Voyages," 15th century.

in Europe. It was drained, as Mr. Galton tells us, of free-thinkers at the rate of 1,000 persons annually, for the three centuries between 1471 and 1781; an average of 100 persons having been executed, and 900 imprisoned every year during that period. The actual data during these 300 years are 32,000 burnt, 17,000 persons burnt in effigy (I presume they mostly died in prison or escaped from Spain), and 271,000 were condemned to various terms of imprisonment and other penalties.* Besides, the celibacy in the Church and in the ecclesiastical institutions checked the transmission of the highest minds and of the highest phases of civilisation. The best men had no families, and the deterioration of the race was the inevitable consequence.

Dr. Döllinger, commenting on the recent "Infallibility of the Pope," gives us his testimony as to the effect of dogma on the general interests of humanity. He says: "As a student of history I cannot accept it, for as such I know that the persistent endeavour to realise this theory of universal dominion has cost Europe rivers of blood, has disordered whole countries and brought them to ruin, has shattered the beautiful structure of the earlier Church, and in the Church of modern times has generated, nurtured, and maintained the worst abuses. As a citizen I must reject it, because by its pretensions to the subjection of States and monarchs and of the whole political system to the Papal power, it leads to endless destructive conflict between Church and State, between clergy and laity."

But the Calvinistic phase of Christianity is fading even in Scotland. J. S. Mill says truly that "what is called Christian, but should rather be termed theological morality, was not the work of Christ or his apostles, but is of much later origin, having been gradually built up by the Catholic Church of the first five centuries, and though not implicitly adopted by moderns and Protestants, has been much less

* Hereditary Genius, p. 350.

modified by them than might have been expected." "The institutions and rites by which Polytheists had testified veneration for their fictitious deities, were now adopted with some slight alterations by Christian teachers, and blended with their worship, on the inadmissible pretext that Pagans would receive Christianity with more facility when they saw the rites and ceremonies to which they were accustomed adopted in the Church, and the same worship paid to Christ and his martyrs which they had paid to their idol deities."*

The fact is, we want a New Reformation. "What little recognition," says Mr. Mill, "the idea of obligation to the public obtains in modern morality, is derived from Greek and Roman sources, not from Christian; as, even in the morality of private life, whatever exists of magnanimity, highmindedness, personal dignity, even the sense of honour, is derived from the purely human, not the religious part of our education, and never could have grown out of a standard of ethics in which the only worth, professedly recognised, is that of obedience."† "Magnanimity, self-reliance, dignity, independence, and, in a word, elevation of character, constituted the Roman idea of perfection; while humility, obedience, gentleness, patience, resignation, are Christian virtues."‡

But old creeds are passing away, and there is even a doubt, as we have seen, among enlightened churchmen, whether the Athanasian Creed shall be retained in the Church services! Some people, not able to accept the anthropomorphic idea of God, reject Him altogether; others, as the Secularists, seeing how much Religion is now made to stand in the way of real progress, reject Religion altogether; and educated people generally, finding Religion and Morality inextricably bound

* Mosheim's Ecclesiastical History, cent. 3, part 2, chap. 3.

† "On Liberty," pp. 88, et seq. ‡ Lecky, vol. 2, pp. 72-155.

together, are everywhere trying to put the "new wine into old bottles," to build the new ideas, as before, in the early centuries of Christianity, on the old foundation. Creeds are seldom now accepted in the literal, but in a non-natural sense; keeping the same words, but giving them a different meaning. *The Pall Mall Gazette*, however, says truly (Mar. 2, 1871): "The sincerity and the amiable desire to reconcile conflicting schools is obvious and praiseworthy; but it results, to speak plainly, in producing such a hash of new theories and old dogmas, such transformations of definite statement into hazy metaphysics, and such unexpected returns to fact from metaphysical theories, that we can find no satisfactory resting place for an ordinary human being." On this Rational Movement in the Church the Rev. J. Llewelyn Davies observes: "If we add to the conclusions of physical science that of historical research, and of the comparative criticism of religions and myths, it is impossible not to feel that an implicit belief in the Old Testament narratives is exposed to most serious difficulties. Any creed, therefore, which is built on the assumption of the perfect authenticity of the earlier Scriptures, is likely to be rudely shaken. * * * The faith taught us by Christian theology is that he whom we know through Christ is the life and order of the universe; and our apprehensions of the mode in which God is related to the universe will rightly vary with our knowledge of the universe itself. It would have been wise for Christians not to tie themselves down to anything more technical and precise on this subject than the broad 'Pantheism' of St. Paul: 'One God and Father of all, who is above all, and through all, and in all.' 'Of him, and through him, and unto him are all things.'"*

But while, on one side, men are turned out of the Church for attacking the doctrines of the Incarnation, the

* "The Debt of Theology to Secular Movements," p. 203. *Contemporary*, January, 1871.

Atonement, and the Inspiration of the Bible, and others still in the Church hold opinions that would seem even to indicate that the very foundations of the faith were giving way beneath us; on the other side we have a solemn judgment (the Purchas case) expected with profound anxiety, that a clergyman must not wear a tunicle, alb, or chasuble in *performing* divine service; he must not mix water with the sacramental wine; he must use bread and not wafers; and he is to stand on the north side of the communion-table and to look south; and by the north side is not to be understood that part of the west which is nearest to the north. These things appear trifling, but the Ritualist really believes that he is performing a miracle when he is consecrating the bread and wine. But this latter injunction that a clergyman *must* stand on the north side, and not at the north end of the west side, has caused quite a rebellion, and we shall probably have a Free Church of England in which the State can take no part. "We see the same effect produced at opposite poles of the Church. The Rationalist tries to prove that he may contradict every Article of the Church so long as he avoids contradicting it in a particular form of words. The Ritualist tries to prove that he may twist every ceremonial to the likeness of the Roman Catholic practice so long as by dexterous interpretations he can keep on the windy side of certain elastic rubrics." Prof. Huxley says: "The clergy are at present divided into three sections: an immense body who are ignorant and speak out, a small proportion who know and are silent, and a minute minority who know and speak according to their knowledge." (*Athenæum*, July 23, 1870.)

Thomas Carlyle, speaking of the effect of creeds in the 17th century, says: "The Christian doctrines which then dwelt alive in every heart, have now in a manner died out of all hearts—very mournful to behold; and are not the guidance of this world any more. * * Nay, worse still, the cant of them does yet dwell alive with us; little doubting that it is cant. * * In which fatal intermediate state, the eternal sacredness of this universe itself, of this human life itself, has fallen dark to the most of us."

We want a new Reformation: the old Religion now "stops the way" and bars the path of progress. Priest-made law has taken the place of Natural law. Impracticable, incredible dogmas have been associated with religious feeling; our veneration and reverence have been connected with certain books, and creeds, and words, and phrases, and formalities, and it is not easy to break the connection; so that religion is *altogether* an exercise of feeling, and not of the intellect—an aspiration and a prayer, and not a logical demonstration. As such it is very difficult to deal with, as it prevents inquiry, narrows the scope of the intellect, and renders error sacred. But we are not without precedent as to the course we must pursue. "Buddha was called omniscient by his earlier pupils; but when in later times it was seen that on several points he had but spoken the language of his age, and he had shared the errors current among his contemporaries with regard to the shape of the earth and the movement of the heavenly bodies, an important concession was made by Buddhist theologians. They limited the meaning of the word 'omniscient,' as applied to Buddha, to a knowledge of the principal doctrines of his system, and concerning these, but these only, they declared him to be infallible. * * * Within the domain of sense and reason, Nâgasina does not claim omniscience and infallibility for Buddha, but he claims for him both in all that is to be perceived by meditation only, or, as we should say, in matters of faith." *

It is to be hoped that Roman Catholics will take the lately-declared dogma of the Infallibillity of the Pope in this sense, and that theologians will consent to leave to the laity the domain of sense and reason, if we make over the domain of faith to them. In the department of Physical Science this

* Max Müller. *Fraser*, May, pp. 589,590.

has already been conceded, and the sun is no longer supposed to go round the earth, and it is admitted that "Nothing in geology bears the smallest resemblance to any part of the Mosaic cosmogony, torture the interpretation to whatever extent we may";* and what we have now to do is to claim the same liberty to extend the domain of sense and reason to mental and moral science. There must be a divorce between religion and morality; they must no longer be proclaimed as necessarily dependent upon each other. There can be no religion without morality, but there may be morality without religion.

Man knows not his whence and whither, and it is to his sacred books that he looks for this revelation; but we must separate metaphor and pure imagination from what we know of fact and reality. Max Müller says: "It is impossible in human language to express abstract ideas except by metaphor, and it is not too much to say that the whole dictionary of ancient religion is made up of metaphors. With us these metaphors are all forgotten. * * The language of antiquity is the language of childhood; aye, and we ourselves, when we try to reach the Infinite by means of mere abstract terms, are but like children trying to place a ladder against the sky." † "Theology," the Rev. J. Llewelyn Davies observes, "has natural tendencies towards corruption. For example, religious enthusiasm gives birth to exaggerated and figurative language; it becomes a point of honour to hand down such language, which is gradually stiffened into propositions; and thus unreality is introduced. Again, nothing is more natural than that men should attribute to God their own modes of thought and feeling, supposing all the while that these have been declared by revelation to be Divine; and it is not long before such corruptions receive the stamp of common acceptance and Church authority.

* "Christianity without Judaism," p. 257, by Rev. Baden Powell.

† "Fourth Lecture on Science and Religion." *Fraser*, July.

Christian tradition, therefore, is no competent guardian of theology." Religious language, although apparently we have forgotten it, can be accepted only as purely metaphorical. For instance, we speak of Heaven and Hell—"and besides these two places for souls separated from their bodies, the Scripture knoweth none"—as above and below,—of ascending into one and descending into the other; but where is above and below in a world that turns round every twenty-four hours? What is above one twelve hours is below the next. Christ is represented as "sitting on the right hand of the Father," and we "cannot be saved" if we do not believe it; but has the Father a right hand? And then, "all the dead shall be raised with the self-same bodies, and none other," although the same bodies must have served half-a-dozen men at least; and there is to be a day of judgment, "in which day not only the apostate angels shall be judged, but likewise all persons that have lived upon earth shall appear before the tribunal of Christ, to give an account of their *thoughts*, *words*, and *deeds*." Certainly this must be a very long day, and it is not easy to believe in it, unless, like Kant, we disbelieve in time and space altogether. We speak of the presence of God in Heaven, and of seeing Him *there* face to face, as if He were not equally present everywhere. We speak of God's sitting on a Throne, surrounded by the Angels of Heaven; of His Majesty, &c., &c., as if our highest ideal were an Eastern Monarch, with all his pomp and state. Surely all this is "the language of childhood," *

* "Religious language is not only the language of childhood, it is the language of the childhood of the race. The theological universe no longer corresponds to that which physical science presents for our contemplation. It was quite different with the Jew. His conception of the abode of Jehovah and the angels, and of departed souls, was exceedingly simple and definite. In the Jewish theory the universe is like a part of a three-storey house. The flat earth rests upon the waters, and under the earth's surface is the land of graves, called Sheol, where after death the souls of all men go, the righteous

but how could any religion that we have at the present day exist without it? There is no religious language that can be

as well as the wicked, for the Jew had not arrived at the doctrine of heaven and hell. The Hebrew Sheol corresponds strictly to the Greek Hades, before the notions of Elysium or Tartarus were added to it—a land peopled with flitting shadows, suffering no torment, but experiencing no pleasure, like those whom Dante met in one of the upper circles of his Inferno. Sheol is the first storey of the cosmic house; the earth is the second. Above the earth is the firmament or sky, which, according to the book of Genesis (chap. v. 6, Hebrew text), is a vast plate hammered out by the gods, and supports a great ocean like that upon which the earth rests. Rain is caused by the opening of little windows or trap-doors in the firmament, through which pours the water of this upper ocean. Upon this water rests the land of heaven, where Jehovah reigns, surrounded by hosts of angels. To this blessed land two only of the human race have ever been admitted—Enoch and Elijah—the latter of whom had ascended in a chariot of fire, and was destined to return to earth as the herald and forerunner of the Messiah. Heaven forms the third storey of the cosmic house. Between the firmament and the earth is the air, which is the habitation of evil demons, ruled by Satan, the "prince of the powers of the air."—("The Jesus of Dogma." *The Modern Thinker*, p. 30.)

Perhaps we have not a more useful and sensible book than "Life: its Nature, Varieties, and Phenomena," by Leo H. Grindon, as long as the author confines himself to what is really known—to observation and experiment; but it is lamentable when he tries to patch his new truth to the above old garment of tradition, which he appears to adopt almost literally. His idea is the common one that God *reigns* apart from nature in heaven, and he tells us that "God desires that all men should be united to him," but he does not tell us how anything can be disunited from the cause of all things, "in whom we live, and move, and have our being;" or how Omnipotence can *desire* anything that it cannot have. Again, he tells us that "the inalienable and irrepressible recollection of the deeds and feelings played forth while in the flesh, provide a beatitude or a misery for ever." (p. 290). Fancy the endless progress of the future being occupied solely with the stupid feelings, deeds, and ideas of the present, and their providing a "beatitude!" We are also told that to *deny* the existence of spiritual substance is to assert that heaven is an empty void, whereas St. John represents it as a plenitude of objects and

used that will bear to be translated literally: still must every man who does not believe in such a human personality as the above be called an Atheist, or *believe himself* to be one! Matthew Arnold truly says "matters are not much mended by taking their (the Scripture writers') language of approximative figures and using it for the language of scientific definition; or by crediting them with our own dubious science, deduced from metaphysical ideas which they never had." * But it is questionable whether the most advanced ideas are in any way applicable to God, inasmuch as they can only be our own attributes, extended beyond limit,—what we call infinitely. Max Müller says that, "closely connected with the belief in a personal God of holiness, is the pursuit of a holy life;" but are we justified in lowering God that we may raise ourselves? He says, "however imperfect and however

scenery, of the most substantial kind. It is to depopulate it also of its angels, who if they be real enough to be *persons*, must assuredly be real enough to consist of substance. Unless always upon the wing, they must likewise have a substantial surface whereon to stand," (p. 121.) Now an American, in showing off his favourite mare, said, we may suppose as the highest praise he could bestow upon her, that she "cocked her tail like an angel;" and whether he had any special knowledge on the subject, or only an intuition of the demands of comparative anatomy, he was more likely to be right than Mr. Grindon, for if angels have wings they must have tails as rudders, and if not "always upon the wing," would require to perch, not stand. Surely it is time that all this was left out of such exceedingly intelligent and useful books as "Life: its Nature, Varieties, and Phenomena," and confined to books of poetry or mythology. There may be worlds greatly superior to our own, and superior intelligences inhabiting them; but to suppose in this universe, where this world of ours is but as a grain of sand, there is some special spot where the Power that sustains all things is specially present, sitting on a throne, surrounded by angels, who are perpetually flying about doing His behests, was all very well for the childhood of the race, but is too childish a conception for the present age. God is equally present *everywhere*, and we must learn to recognise Him, or He will be recognised, as is at present the case with too many, *nowhere*.

* "Literature and Dogma," p. 46, *Cornhill*, July, 1871.

childish the conception of God may be, it always represents the highest ideal of perfection which the human soul, for the time being, can reach and grasp. Religion therefore places the human soul in the presence of its highest ideal; it lifts it above the level of ordinary goodness, and produces at least a yearning after a higher and better life—a life in the light of God."* We may fairly ask whether, as Religion is now understood, this is not placing *our* ideal a little too high, and whether it does not lift "the level of ordinary goodness" too much above this world altogether? In the New Reformation religious duty will assume quite a new aspect. Let us see what light we have to guide us on the subject. What do we know of God?

To the Physicist and Comtist there is nothing in the world but Force, Matter, and Motion—or its modes of action. "Every form is force visible; a form of rest is a balance of forces; a form undergoing change is the predominance of one over others." But this force must be the force or power of something; this something is unknown to us but as the Source of all power, and therefore the cause of everything—the Theist calls it God. Although not known to us, however, in its essence, it is partially known to us in its Attributes and Manifestations. We speak of the power of God, as if it could be delegated or separated from Him; but to separate power or force from that of which it is the force, is clearly as impossible and inconceivable as to separate motion from the thing moving. This "Unknown" entity is the "Substance" of Spinoza; and the "Being" that underlies all forms of Being, of Hegel; and in creation, as Hegel says, "Being—Becomes." Mind and Matter are only phenomenal modifications of this common "substance." "There is but one infinite Substance," says Spinoza, "and

* *Fraser*, July, p. 107.

that is God. He is the universal Being, of which all things are the manifestations. * * * Extension and Thought are the objective and subjective, of which God is the Identity." We know nothing but our own Consciousness, and the different "modes of motion" which we infer (or are conscious of) are the cause of it. We know nothing outside this consciousness—that is, we know nothing of the universe but as "a manifestation of Power from something that is *not* I" (Fichte). The classification and arrangement of these "modes of motion," of these manifestations of Power, we call Science. A change of position, a change of form, a correlative of force, is all that is going on, and the difference between chaos and order, between barbarism and civilisation, is simply a difference of arrangement. Now it is with the Cause of this change, with the Power that affects it, with the tendency of the arrangement, that Religion has to deal; and we necessarily come to the same conclusion as Saint Paul, that "There are diversities of operation, but the same God worketh all in all." "When we view the world as one universal effect, we are at once led to the contemplation of a universal Divine Agency. Does not the Infinite act on every atom? God never delegates His power; He cannot transfer divinity to a substance: there is no power, therefore, separate from Himself. In Him all things have their being."*

This idea is much more purely and perfectly expressed in some of the Eastern religions; and apart from the abuses that were afterwards grafted upon it, it is without that material, anthropomorphic element which distinguishes the religions of our Western nations. Mr. W. R. Alger, in his notice of Buddha, says: "To stigmatise such a man, in the opprobrious sense of the words, as an atheistic eulogiser of nothingness, a godless unbeliever, is manifestly unjust.

* "The Philosophy of the Bible," pp. 35-40, by Rev. J. White Malller, M.A.

Absolute pure being is nothing definite, is no *thing.* It is All. As Spinoza, with other metaphysical masters before and after him, has said, every determination of being is a negation; every attribute or quality affirmed of it is a limitation. Now Gotama Buddha's doctrine of the extinction of existence means the removal of limitations, the destruction of all obstacles to the return into that pure being, whereof, as indicated by the word Nirwána, he himself says: We can affirm nothing, neither that it is nor that it is not, since it has no qualities. It is that conditionless state, the idea of which it bewilders the faculties of thought to conceive, and baffles the resources of language to express; although the writings of every deep speculative philosopher, from Heráclitus to Hamilton, deal familiarly with it. The scientific idea of force is the idea of as pure and mysterious a unity as the One of Pármenides. It is a noumenal integer phenomenally differentiated into the glittering universe of things. The Christian who asserts that the unknowable cause of all is an intelligent and affectionate Father, a personal counterpart of himself dilated to immensity, would brand as an atheistic nothingarian the scientist who pauses with the idea of a unit of force, and denies substantive validity to everything else. And yet to the philosopher who has adequately thought his way to that conception, with the fit emotion, it is unquestionably a conception of overwhelming religiousness, capable of yielding an unsurpassed measure of authority and trust, of awe, sweetness, and peace. * * The perception of the indivisible unity of real being, and the purely phenomenal nature of the self—in the faith of Buddha this is the matchless diamond whose discovery sets every prepared slave free. * * Like Hume, Spinoza, and other subtle masters of thinking, Buddha fancied he saw the delusiveness of all selfhood; saw that the soul is no substantive unit, but merely a current of states; its sole identity consisting of the accumulated mass of associations in experience, the organic conditions of memory."* Mr. Alger tell us that "the system spread rapidly over one nation after another, drawing swarms of converts, until it became, as it is at this moment, after the lapse of twenty-five hundred years, the most numerously followed of all religions that have ever prevailed on the earth. The ignorant myriads of his followers, unable to understand or be satisfied with the transcendent abstractions of the system, deformed its teachings by the addition of their superstitious notions, and ended, in many cases, by deifying the sage himself, and painting a new paradise in the abyss. Still, in all its forms, the religion retains much of the metaphysical speculation, and more of the sublime ethics of its founder."† "Buddha is One, the One; and it is only with the inward eye, purged from sensual corruption, and steadily fixed on the contemplation of unity, that he can in anywise be apprehended. For the natural eye of the ordinary man views a multiplicity of things, each thing separate and divided from the other. The natural eye takes account only of appearances; it requires the severest discipline for a man to behold the Reality. * * The one infallible diagnostic of Buddhism is the belief in the infinite capacity of the human mind. * * Lastly, the idea of deified man *is there;* but this loses itself in another, that there is in man, in humanity, a certain Divine Intelligence, which at different times, and in different places, manifests itself more or less completely, and which must have some central manifestation."‡ It is curious how some of the sayings and parables of Buddha resemble those attributed to Christ, thus: "Let a man overcome anger by love, evil by good, the greedy by liberality, the liar by truth."

"What is the use of platted hair, O fool? What of the

* "The Genius of Solitude," pp, 201, 202, 191. † Ibid. p. 200.

‡ "The Religions of the World, and their Relation to Christianity," by Rev. F. D. Maurice, pp. 76-85.

raiment of goatskins? Within thee there is ravening, but the outside thou makest clean,"* and many others.

We have the same idea in the Wahhabee or purist Doctrine of Mohammedanism. "La Ilah illa Allah; there is no God but God; their full sense is, not only to deny absolutely and unreservedly all plurality of nature or of person in the Supreme Being, not only to establish the unity of the Unbegetting and Unbegot, in all its simple and uncommunicable Oneness, but besides this the words, in Arabic and among Arabs, imply that this one Supreme Being is also the only Agent, the only Force, the only Act existing throughout the universe, and leave to all beings else, matter or spirit, instinct or intelligence, spiritual or moral, nothing but pure unconditional passiveness, alike in movement or in quiescence, in action or in capacity. The sole power, the sole motor, movement, energy, and deed is God; the rest is downright inertia, and mere instrumentality, from the highest archangel down to the simplest atom of creation." This Mr. W. G. Palgrave calls the Pantheism of Force, and certainly it is the scientific idea run to its greatest abuse in Eastern Fatalism, and carried on, as we find, to Calvinistic fatalism. "This Force or Power is supposed to be exercised according to the capricious Will of a human Autocrat, being especially jealous of his creatures, lest they should attribute to themselves something of what is His alone. When God, so runs the tradition, resolved to create the human race, He took into His hand a mass of earth, the same whence all mankind were to be formed, and in which they after a manner pre-existed; and having then divided the clod into two equal portions, He threw the one-half into Hell, saying, 'These to eternal fire, and I care not;' and projected the other into Heaven, adding, 'and these to Paradise, and I care not.'"†

* Palgrave's "Central and Eastern Arabia," vol. 1, p. 366.

† "Science of Religion," *Fraser*, July, 1870, p. 106.

This is the Calvinistic doctrine in its simplest and most merciful form; for with the Calvinist, broad is the road that leadeth to destruction, and narrow the path of life, and all has been preordained by an eternal and immutable decree, not from mere indifference, but "out of his mere love, for the praise of his glorious grace." Thus, we are told by the Larger Catechism of the Church of Scotland, established by Act of Parliament, and which the great majority of Scotch children are taught to venerate as an unquestionably correct interpretation of Scripture, that "God, by an eternal and immutable decree, out of his mere love, for the praise of his glorious grace, to be manifested in due time, hath selected some angels to glory; and, in Christ, hath chosen some men to eternal life, and the means thereof: and, also, according to his sovereign power, and the unsearchable counsel of his own will (whereby he extendeth or withholdeth favour as he pleaseth) hath passed by and preordained the rest to dishonour and wrath, to be for their sin inflicted, to the praise of the glory of his justice."

The Science of the present day recognises but one Force, and its Correlations, or transformations and modes of action. What is this the Force of? Unknown, says the man of Science, except in these modes of action. It is the Force of Matter, says the Materialist; of Mind, says the Spiritualist; but Mind and Matter are but "phenomenal modifications of the same common substance." But what is this common substance? "There is but one infinite Substance," says Spinoza, "and that is God," and so say all the great religions of the world. The one infinite Substance and the Force of our modern scientist is the same, and the question is, what do we and can we know of it? It is known to us in its unity, its manifestations or modes of action, and in its purpose or design.

Science proves the Unity of Force; for throughout the

Universe, as far as observation extends, every atom is pulling at every-other atom, and that with set purpose, which proves both intelligence and volition. "Their movements, their interchanges, their 'hates and loves,' their 'attractions and repulsions,' their 'correlations,' their what not, are all determined on the very instant. There is no hesitation, no blundering, no trial and error. * * The presence of MIND is what solves the whole difficulty; so far, at least, as it brings it within the sphere of our consciousness, and into conformity with our own experience of *what action is*."* But this intelligent action, as far as we know, is not now attended with consciousness, but has passed out of its sphere, and how can we account for that? "The necessity may be disguised, but can never be escaped, of interpreting the universe by Man." In man a voluntary action, often repeated, becomes involuntary, or automatic—that is, is performed unconsciously. We have thus intelligent action without consciousness; and "in conformity with our own experience of what action is," all force or power is Will power—the will of God. The will which originally required a distinct *conscious* volition has passed, in the ages, into the unconscious or automatic, constituting the fixed laws and order of nature.† Natural

* Sir J. F. W. Herschel.

† "Causation is the will, Creation the act of God." W. R. Grove.

"In putting forth our Will (using the term for the whole activity that may become voluntary), we certainly know the self as *Force;* we get behind the phenomena which we produce, and are let into the secret of their origin in a way which we should miss if we only looked upon them. In other words, we know ourselves as *Cause* of them. It is then under the form of Will that we are introduced to Causality; and the axiom resolves itself into the proposition, Every phenomena springs from a Will," &c.—("Is there any Axiom of Causality?" by Rev. James Martineau. *Contemporary Review*, July, 1870.)

"1. *Matter*, as an entity distinct from forces which alone we perceive, cannot be proved to exist, and is beside inconceivable, and utterly unnecessary.

Causation, therefore, and God's Will are the same, and cannot be separated. This Will does not in any case, as far as we know, act by way of special creation, but by the "law" or order by which everything is made to make itself—the present dependent upon the past, the future dependent upon the present. Everything necessarily acts according to its nature, and it is the Creator's nature always to act in the same way. It is ignorance and uncertainty only that change and vary; hence harmonious order and fixed laws, or what we call "Harmonious order governing eternally continuous progress." "A phenomenon is explained," says Huxley, "when it is shown to be a case of some general law of Nature; but the supernatural interposition of the Creator can, by the nature of the case, exemplify no law." If I might venture to differ from so high an authority, I should say, yes, it can; it can exemplify a higher law, the law of the Creator's own nature, not acting blindly, but intelligently; whereas, what we call "a general law of Nature" is where the Creator's intelligent and intellectual attributes have passed into unconscious or automatic action.

Man is a Cosmos—the whole world exists in him in miniature; and judging great things by small, so far as we can understand him, so far can we understand how the world has been made and is governed. *

"2. *Force* is the one thing our senses are cognisant of, and therefore the one thing our reason has any foundation for believing in.

"3. We are *conscious* of exerting *force*, while we have no knowledge that *matter* (which is force) has any power to exert force.

"4. The *inference* is, that *force* is produced in the only way we know force to be produced, by the will of conscious beings."—(A. R. Wallace.)

* "To know ourselves is to know God, for we are a similitude of the Deity—a living image of the eternal divine nature. That which is the triune God is manifested in nature and creation; and of this entire nature and creation man is the epitome."—(Jacob Böhme.)

"The human body: that wonderful machine, the richest, the most

The world began in a nebulous mist: so man began in protoplasm—the nebulous matter of life. Each living function, beginning with an animal all stomach, at first was probably a voluntary action. As part was added to part, and function to function, in different forms of life, and as these functions descended to one animal after another, and rose in the scale of being, they gradually lost their voluntary character and became automatic, as the functions of brain do now when they pass into instincts. The action of the stomach, heart, lungs, &c., at first voluntary, by frequent repetition became involuntary; the action of the brain and nervous system, also, upon which intelligence depends, passed, in many cases, from voluntary to involuntary, and in automatic action we have intelligent action without consciousness. Every atom in the body performs its action intelligently yet unconsciously—even the brain itself when healthy, for we have no consciousness that the action of the brain is necessary to thought. It is curious and instructive to watch, both in the child and in the man, the stages by which the conscious gradually passes into the unconscious—the voluntary and volitional into the automatic, in walking, winking, talking, touching, seeing, writing, &c. Co-ordination of mind and movement is at first effected with difficulty, and requires a mental effort, and then becomes so easy that it has a tendency even to usurp

diversified, the most complete of all, yea, that masterpiece of nature which contains all the forms of combination and organisation which nature can produce; that little world in which is reflected the entire universe. The whole of nature is one individual. The parts vary infinitely, but the individual in its totality undergoes no change."—(Spinoza.)

"External nature has a body and soul like man, but the soul is the Deity."—(Ruskin.)

"If there be any work in Nature which reflects any image of the Creator, the human mind is that work. And the highest light which these faculties supply is that by which mind recognises in Nature the working of a spirit like its own."—("On Variety as an Aim in Nature," by the Duke of Argyll. *Contemporary Review*, May, 1871.)

the action of the mind.* So what we call the Laws of Nature were probably at first *voluntary* action of Divine Will power, but are now no longer attended with consciousness. In man this liberated consciousness is required for higher purposes, or enjoyment; and if we trace progress in Nature, it must exist in God its Author also. Force in this way becomes blind, and some of those apparent exceptions to design occur upon which certain philosophers ground so much. Organs that were useful lower down in the scale are sometimes passed on to higher grades, where they are no longer of any use. Man's rudimentary tail, the *os coccyx*, for instance, is not now of much use to him, whatever it may have been to his possible progenitor—the monkey. The guinea-pig, Huxley tells us, has teeth which are shed before it is born, and hence can never subserve the masticatory purpose for which they seem contrived; and in like manner the female dugong has tusks which never cut the gum.† We have animals that never swim, and yet their fingers are provided with the requisite membranous apparatus, and numberless other monstrosities and structures, which, so far as we can judge, are neither injurious nor beneficial, but which furnish, in my opinion, illustrations of the automatic character of the voluntary, conscious Will power, by which they were originally brought about. It would seem as if Nature, having got into a habit (automatic action) of making teeth, and so of other organs, makes teeth sometimes, in the guinea-pig for instance, when they are not wanted: we have the continuance of habitual actions when the motive for them has ceased. But is not this exactly what we might expect from our knowledge and experience of the action of the little spark of mind in ourselves which has been parcelled off from the infinite whole?

The Unknown, the one Infinite Substance—that which underlies all Force and all phenomena—is known to us in

* See Dr. W. B. Carpenter's Article in *Contemporary Review*, May, 1871.

† "Lay Sermons," p. 284.

its manifestations or modes of action; all Power being thus Will power, conscious or automatic. All Science is a mere classification of a difference of arrangement and modes of action, and the real source of this action is hidden in what are called secondary causes, when in fact there are no secondary causes—unless we designate conscious and what appears to us as unconscious volition as primary and secondary cause. All is immediately from God. "And this immediate action of God is not limited to the mind of man. It is the same through all nature. God has not given up his creation to secondary causes. God never leaves His world. He is present in it now as much as in the first moment of creation; in fact, creation never ceases. The same will, the same power, and the same presence that were required to create the world, are required every moment to preserve it. What we call the laws of nature are but the expressions of the will of God. He works by laws, but the working is not therefore less immediate or less dependent on His will and power." * "A human mechanist may leave the machine he has constructed to work without his further personal superintendence, because when he leaves it God's laws take it up; but when God has constructed his machine of the universe he cannot so leave it, or any the minutest part of it, in its immensity and intricacy of movement, to itself, for if He retire there is no second God to take care of this machine. Not from a single atom of matter can he who made it for a moment withdraw His superintendence and support; each successive moment all over the world, the act of creation must be repeated." †

"There is but one infinite Substance, and that is God." There is but one Force of that Substance, and Heat, Light, Magnetism, Electricity, Attraction, Repulsion, Chemical Affinity, Life, Mind or Sentience, are the names by which we distinguish the manifestations or different modes of action of this

* Malebranche.

† "An Essay on Pantheism," p. 212, by Rev. John Hunt.

Force. This Substance or Force is constantly passing into new forms, the old forms ceasing to exist. This sequence or correlation we call cause and effect, and each cause and effect is a new Life, a new Death: each form being a new creation, which dies and passes away, never to return; for nothing repeats itself. "All Nature and Life are but one *garment*, a 'living garment,' woven and ever a-weaving in the Loom of Time."* God and Nature are inseparable. God *is* Nature, not *in* Nature. The Laws of Nature are the Attributes of Deity, and all action His consequent and necessary mode of working.†

* Sartor Resartus, p. 210, T. Carlyle.

† Of course these conclusions may not be accepted, and may not be even comprehensible, but by those who are acquainted with the last results in Physics and Metaphysics. We know only our own consciousness, which we call spiritual, and the world is created in that consciousness by a manifestation of Power or Force from something which is not I, but which power or force is equally spiritual. *Ex nihilo nihil fit*, the world cannot have been made out of nothing; it must therefore be an Evolution from God, or rather in God. The common notions on the subject are the delusions of the senses, and of our intellectual faculties or modes of thought.

"Evolution notions are absurd, monstrous, and fit only for the intellectual gibbet in relation to the ideas concerning matter which were drilled into us when young. Spirit and matter have ever been presented to us in the rudest contrast, the one as all noble, the other as all vile. But is this correct? Does it represent what our mightiest spiritual teacher would call the Eternal Fact of the Universe? Upon the answer to this question all depends. Supposing, instead of having the foregoing antithesis of spirit and matter presented to our youthful minds, we had been taught to regard them as equally worthy and equally wonderful; to consider them, in fact, as two opposite faces of the self-same mystery. Supposing that in youth we had been impregnated with the notion of the poet Goethe, instead of the notion of the poet Young, looking at matter, not as brute matter, but as 'the living garment of God,' do you not think that under these altered circumstances the law of Relativity might have had an outcome different from its present one? Is it not probable that our repugnance to the idea of primeval union between spirit and matter might be consider-

What, then, is the necessary inference? The Eternal Sacredness of the Universe, and of all Life in it, which, as Carlyle says, "has fallen dark to the most of us."

Religion must be brought into harmony with Philosophical Necessity. All is God, and therefore all is Good. Good and Evil are purely subjective—that is, the differences we make in this respect are of our own creation, and exist only in ourselves. They indicate only the different way in which we are affected by the manifestations of power without us, and to which it is our first religious duty to adapt ourselves. They are the mental springs of action—the pleasure and pain, that is, the good and evil being at present equally necessary, although as we progress we shall less and less require the evil. As Shakespeare says, "Virtue itself turns vice, being misapplied," or to put it differently, vice is only virtue in the wrong place. All the sets of laws, Physical, Moral, and Intellectual, are equally sacred, spiritual, and to be obeyed. "Nature is conquered by obedience" only, and by knowledge and obedience are we enabled to choose the good and refuse the evil.

Everthing has received a definite constitution, which is unalterable by us; but knowing what this constitution is, we can regulate our conduct accordingly. If we put ourselves in opposition to this fixed order of things we are punished—that is, are made to suffer for it; but this punishment is not vindictive or revengeful, but for our good. All pain or punishment in Nature is for the good or reformation of the person suffering, and no one, either in this world or another, would wish to be "saved" from that which was for his own good. Forgiveness of sin would be an injury. And this shows us the real nature of that much-abused word, "Responsibility."

ably abated? Without this total revolution of the notions now prevalent, the Evolution hypothesis must stand condemned; but in many profoundly thoughtful minds such a revolution has already taken place."—(Prof. Tyndall on the "Scientific Use of the Imagination.")

Responsibility does not mean the vengeance to which we are liable for the past, which cannot now be altered, and, indeed, could not at any time have been different under the existing circumstances. The same causes must always again produce the same effects. Responsibility simply means that we must always bear the natural and necessary consequences of our actions. From these we can never get away. But as the order of nature is fixed, as we say—according to Law, we can in most cases by study know what these consequences are, and by altering our conduct avoid them where they are disagreeable. This is the standard by which all opinions must be tested—viz., the results which they produce when reduced to practice.

The Law of Consequences is the Law by which the world is everywhere governed. It is the Revelation given to all without respect of persons. But this Revelation has been neglected by Divines in favour of the horrid, inhuman dogmas of the dark ages: they prefer rather to tell us what is going on, and what is to go on, in other worlds rather than in this. God, they say, is only to be found in heaven, not here. But the rightly-tutored mind sees God everywhere and in everything. "There is no other revelation than the ever continuing."* All is sacred, and the proper function of Religion is to trace the *good* in everything—to trace all things to God; to try to discover His mode of working, and what the experience of probably millions of years has found to be the best and fittest, and which a "Natural Selection" has established.

But man does not thus humbly wait upon God, but makes a world of his own, from which all the blunders he thinks he recognises here shall be omitted; and this brings us to the third way in which the Great Unknown is known to us—viz., the "purpose" in His manifold "manifestations and modes of action."

* Jean Paul Richter to his Son.

Although it is impossible to conceive a "beginning," yet the present order of things had a beginning, and "matter assumed a definite motion at a definite time." By a definite motion we mean a motion in one particular direction rather than in another; and the question is why it began to move, and for what purpose.* As we have already seen, the purpose we find in creation, the object of the creative Principle, is the production of the largest amount of pleasurable con-

* "I had rather believe all the fables in the legend, and the Talmud, and the Alcoran, than that this universal frame is without a mind."—(Bacon, Essays, 17.)

"The whole frame of Nature bespeaks an intelligent author; and no intelligent inquirer can, after serious reflection, suspend his belief a moment with regard to the primary principles of genuine Theism and Religion."—(D. Hume, "Dialogues concerning Natural Religion," part 10.) Hume, however, rejected with scorn the appeal to the solution which another world was to give to the difficulties of this, which he designated as "building on air, and establishing one hypothesis upon another."—(Ibid.)

"The birth both of species and of the individual are equally parts of that grand sequence of events which our minds refuse to accept as the result of blind chance. The understanding revolts at such a conclusion."—("Descent of Man," part 2, p. 396.)

"If we cannot believe in the relations which He has established between the Mind of Man and the rest of His creation, we can believe in nothing. We are ourselves 'magnetic mockeries' in a world of lies."—(Duke of Argyll, *Contemporary Review*, May, 1871.)

"Overpoweringly strong proofs of intelligent and benevolent design lie all round us, and if ever perplexities, whether metaphysical or scientific, turn us away for a time, they come back upon us with irresistible force."—(Sir W. Thomson's Inaugural Address at the British Association, Edinburgh, August, 1871.)

"Nor will men easily loosen from their errors, and enter the temple of nature, and of the God of nature, which is, that infinite cause in nature, eternal, omnipresent and without change—the principle of matter and of the properties of matter, motion, and the mind of matter, but neither matter, nor property, nor mind. What it is, is beyond our comprehension, and folly to suppose. The finite cannot grasp the infinite, nor phenomena a cause.—("Letters on the Laws of Man's Nature and Development," p. 343, by H. G. Atkinson.)

sciousness. All unconscious action is, as we have seen, the effect of automatic Will power; so that a previous conscious action must have underlaid this unconscious, and probably afterwards passes through certain "modes of motion" from the automatic to a higher state of pleasurable consciousness. The "individuality" by which this takes place is created in our own minds; unity must be behind it. This, it may be said, is making the nature of God progressive. Certainly I do not see how we are to resist the inference. If the World is progressive, if the Universe is progressive, God *is* the Universe. The opposite idea, that God is separate from the Universe, is derived from our experience of our own creations, which work without our personal power or aid, forgetting, as we have previously seen, that we have borrowed the power by which they work from the Source of all power. Besides, there cannot be the world *and* God: that would be a limitation of the Infinite. The author of "The Pilgrim and the Shrine" truly says (vol. 2, p. 18): "Had absolute perfection existed prior to the 'creation' the universe had never been. For why create, unless to produce a state of things better than before existed? For the greater Glory of God? Then His Glory was capable of increase, and therefore was not complete. Therefore there was no perfection; and therefore no God! Into this maze of contradiction do the theologians lead us, by detaching God from the universe, and making Him a thing apart."

God's supposed "Foreknowledge" presents a much greater and more incomprehensible difficulty than the supposition that He is one with the Evolutionary Progress that we see. Infinite Foreknowledge implies that everything must come to pass as it has been foreseen it will. As Jonathan Edwards says: "There is as much of an impossibility but that the things that are infallibly foreknown should be, (or which is the same thing), as great a necessity of their future existence as if the event were already written down, and was known and read, by all mankind through all preceding ages, and there was the most indissoluble and perfect connection between the writing and the thing written. In such a case it would be as impossible the event should fail of existence, as if it had existed already; and a decree cannot make an event surer or more necessary than this."* Everything with God, therefore, would be the same as if it had already happened, with no possibility of change. Surely this is the deadest fatalism, although it is certainly no more than saying that what will be will be, whether it is known what it is that will be or not. But then it is the *knowing*. "Surely rather than this," says the author of "The Pilgrim and the Shrine," "it is easier to believe that the existing series of things is the actual first thought or life of God; * * otherwise to the infinite foreknowledge all is still certainty: there is no trust and no hope." "One everlasting Now" sounds very fine, only it is a little inconceivable. The Will and every mental state grows out of what we call physical power or force, and obeys its laws; and it is most probable therefore that the "ego," the "individuality," is a delusion—a mere creation of the mind, and all happiness and joy is a unity, like the force of which it is the correlate.

Alphonso the Wise, of Castile, constructed the Alphonsine Tables on the Ptolemaic hypothesis that the sun went round the earth, then universally received; but he suspected there was something wrong, and ventured to suggest that if the Creator had consulted him, he should have recommended a solar system far more simple and beautiful, with the sun in the centre, etc. So theologians now place man in the centre of their system, making everything to revolve round him, and when they find things do not square with this hypothesis, they think the world a very bungling contrivance, and make another on the pattern of that which they think this ought to have been if it had not been spoilt. Man by his pride

* "Inquiry concerning the Freedom of the Will."

and self-conceit makes it dark, and then complains that he cannot see. It is true this world to our ignorance is "a mighty maze, but not without a plan." Man is but a small part of a much larger system, even on this planet, and is moved—notwithstanding the ridiculous airs he assumes, and all the fine names and supposed differences by which he distinguishes himself—by the same laws as all the rest of the sensitive creation. The world—supposed to be made for him alone—was inhabited hundreds of thousands of years before he came upon it, and at this time there is a whole Ocean World, and world *within* world besides, of sensitive existences in which he can take no part. The phenomenal world is not a reality: each creature creates its own world and carries it about in its own head; outside itself there is nothing but the play of forces on the nervous centres of each being. This creates a thousand worlds adapted to the different wants of each, and is so much better, therefore, than if it had objective reality; and man stupidly thinks it was all made for him, and that the way these forces affect him is the absolute measure of truth: whereas a very few only of the modes of action of the forces around him reach him through his five senses or otherwise, and produce those impressions and ideas which are sufficient to guide him towards the object of his being, towards his *real* world—that of his pleasure and pain, and which he calls his moral world. In reality he knows very little indeed of all that is going on without him, and he is cognisant only of the influences that can penetrate through his thick skin; while whole worlds of beings may exist outside his intellectual ken.

The "plan" being to create the largest amount of enjoyment in all these worlds of varied sensibility, the mode of proceeding is altogether different to the rule or laws which men lay down for the regulation of their intercourse with each other, and which they call moral laws. God proceeds upon a principle of Natural Selection, by which only the strongest and fittest and those most capable of enjoyment are preserved, and there are as many of them always preserved as there is room for. The weak are evidently in the way, and are squeezed out of existence—not, however, without some little outcry on their part, and the making of another world in which more *justice* is to be done to them. It is evident, however, that it is only *selfishness* that makes them stand in the way. Nature cares nothing for individuals:

So careful of the type she seems,
So careless of the single life.*

* LIV.

The wish, that of the living whole
No life may fail beyond the grave,
Derives it not from what we have
The likest God within the soul?

Are God and Nature then at strife,
That Nature lends such evil dreams?
So careful of the type she seems,
So careless of the single life;

That I, considering everywhere
Her secret meaning in her deeds,
And finding that of fifty seeds
She often brings but one to bear,

I falter where I firmly trod,
And falling with my weight of cares
Upon the great world's altar-stairs
That slope through darkness up to God,

I stretch lame hands of faith, and grope,
And gather dust and chaff, and call
To what I feel is Lord of all,
And faintly trust the larger hope.

LV.

"So careful of the type"? but no.
From scarped cliff and quarried stone
She cries 'a thousand types are gone:
I care for nothing, all shall go.

The whole surface of the earth is one network of nerves quivering with enjoyment, so greatly exceeding the pains that it is probable that in the aggregate they are not perceived at all. Life is thus always preserved at a high pressure. "There is no exception to the rule," says Darwin, "that every organic being naturally increases at so high a rate, that if not destroyed this earth would soon be covered by the progeny of a single pair."

It is this *necessary* destruction—by which the fittest only are preserved, and life ever kept at a high pressure and the greatest possible capability of enjoyment—that man finds

'Thou makest thine appeal to me:
I bring to life, I bring to death:
The spirit does not mean the breath:
I know no more.' And he, shall he,

Man, her last work, who seem'd so fair,
Such splendid purpose in his eyes,
Who roll'd the psalm to wintry skies,
Who built him fanes of fruitless prayer,

Who trusted God was love indeed,
And love Creation's final law—
Tho' Nature, red in tooth and claw
With ravine, shriek'd against his creed—

Who loved, who suffer'd countless ills,
Who battled for the True, the Just,
Be blown about the desert dust,
Or sealed within the iron hills?

No More? A monster then, a dream,
A discord. Dragons of the prime,
That tare each other in their slime,
Were mellow music match'd with him.

O life as futile, then, as frail!
O for thy voice to soothe and bless!
What hope of answer, or redress?
Behind the veil, behind the veil.

—Tennyson's "In Memoriam."

so difficult to reconcile to the character he has formed of God. He

> Who trusted God was love indeed,
> And love Creation's final law—
> Tho' Nature, red in tooth and claw
> With ravine, shriek'd against his creed.

It is evident his creed is a wrong one, and that if he will reconcile his belief with the actual phenomena of the world he must alter it. In fact, the terms which we have been in the habit of applying to God, such as holiness, purity, truth, justice, and love, have no meaning as applied to the creative Principle They apply to man alone, and to his relations with his fellow-creatures, and are no more properly applicable to God than to give Him a man's bodily form, and to speak of His head and arm, and of His sitting upon a throne. They are man's mental form. The Universe is God's form; and our Religious duty is, by better knowledge of it, to be able to see in it that which is really worthy of Him. The stars of earth are quite as wonderful and as beautiful as those of heaven, but the darkness that reveals the one obscures the other; the mental darkness of open day being as great as night itself.* He who sees

* Baron Bunsen, in his "God in History; or, the Progress of Man's Faith in the Moral Order of the World," says: "The God of the Israelitish and Christian Judaisers, from Ezra to Moses Mendelssohn, is the so-called 'personal God' or 'Supreme Being,' consequently *a* being, like other beings. They may contrive as many fine phrases as they will to designate God as the Spirit; their God is still, only in the highest conceivable sphere, a venerable, corporeal patriarch, who dwells outside the universe, though He is called the all-pervading Spirit; severed by an impassable gulf from the universe and from the human spirit which he is yet said to inhabit; not distinguished from the universe, as the Infinite from the Finite, but as mutual exclusives, like the watchmaker and the watch; out of space, and therefore banished from the universe, or else dwelling in a space by himself, as the Soul does, according to the view of certain physiologists. Such a God can have no other cultus than an external and

not God "face to face" in this world is not likely to have any clearer insight in another. The "splendour in the grass, the glory in the flower," does it belong to the grass or flower, or to God? Tennyson himself does not leave us without one look "behind the veil." Thus he says:

> And forth into the fields I went,
> And Nature's living motion lent
> The pulse of hope to discontent.
>
> I wondered at the beauteous hours,
> The slow result of winter showers:
> You scarce could see the grass for flowers.
>
> I wonder'd, while I paced along:
> The woods were filled so full with song,
> There seemed no room for sense of wrong.
>
> So variously seem'd all things wrought,
> I marvell'd how the mind was brought
> To anchor by one gloomy thought;
>
> And wherefore rather I made choice
> To commune with that barren voice,
> Than him that said, "Rejoice, Rejoice!"

ritualistic; the truly ethical element in worship is as much repressed and driven into the background by the externality of ceremonialism as by the breaking up and materialising of our conception of Him whom the conscience recognises to be One and Sole. Nay, that spurious monotheism is in some respects still further removed from living faith; namely, in so far as it betokens a narrowing of the highest apprehension of God, the Eternal. Up to the time of Ezra, the safeguard of the Mosaic conception of God lay in the spirit animating the Law and the Prophets, and the testimony which these bore against ceremonialism and justification by works, against priest-craft and tyranny, as they did in later times Pharisaism and Sadduceeism; consequently, in the spirit of the written, traditional, historical word of God, proclaiming the mighty acts of Jehovah among the children of men. Now, in every age, such a prophetic guild and such documentary testimony had been wanting to the Greeks; and, as we have seen, they lost nothing in this respect by not erecting the Orphic hymns and Sibylline oracles into a word of God, but rather allowing scope for a free intercourse of the living mind with the Deity."

Lord Palmerston, in 1854, was applied to by the Presbytery of Edinburgh to be informed whether he intended, as Home Secretary, to advise the Queen to order a day of fasting, humiliation, and prayer, to be held in Scotland, in order to supplicate Divine Providence to stay the cholera which then afflicted the people. His reply is memorable as a step in high quarters towards the initiation of the New Reformation. He said:

"The Maker of the universe has established certain laws of nature for the planet in which we live, and the weal or woe of mankind depends upon the observance or the neglect of those laws. One of those laws connects health with the absence of those gaseous exhalations which proceed from over-crowded human beings, or from decomposing substances, whether animal or vegetable; and those same laws render sickness the almost inevitable consequence of exposure to those noxious influences. But it has at the same time pleased Providence to place it within the power of man to make such arrangements as will prevent or disperse such exhalations, so as to render them harmless; and it is the duty of man to attend to those laws of nature, and to exert the faculties which Providence has thus given to man for his own welfare.

"The recent visitation of cholera, which has for the moment been mercifully checked, is an awful warning to the people of this realm, that they have too much neglected their duty in this respect, and that those persons with whom it rested to purify towns and cities, and to prevent or remove the causes of disease, have not been sufficiently active in regard to such matters. Lord Palmerston would, therefore, suggest that the best course the people of this country can pursue to deserve that the further progress of the cholera should be stayed, will be to employ the interval that will elapse between the present time and the beginning of next spring in planning and executing measures by which those portions of the towns and cities which are inhabited by the poorest classes, and which, from the nature of things, must most need purification and improvement, may be freed from those causes and sources of contagion, which, if allowed to remain, will infallibly breed pestilence, and be fruitful in death, in spite of all the prayers and fastings of a united but inactive nation. When man has done his utmost for his own safety, then is the time to invoke the blessings of Heaven to give effect to his exertions."

Our religious duty thus lies, not in preparing ourselves for *another* world, but it lies very often under our very nose in this, and our nose points out the direction in which it lies, and we have only to follow it. The majority of the Presbytery were much scandalised by being told this by Lord Palmerston, and doubtless would have preferred prayer and fasting as much easier, if not so efficacious; but, nevertheless, the civic rulers of Edinburgh acted upon his lordship's advice with very beneficial effect.

But what of the efficacy of prayer and fasting? The exercise of the reason of man, no less than the instinct of the brute, are dependent upon the "order of nature" being to-day as it was yesterday; for reason infers that which will be from what has been, and the delicate relationship between instinct and object, which it has taken ages to establish, must be maintained, or animal life must cease to exist. All depends upon the uniformity of Nature's law, and this uniformity cannot be interfered with at man's intercession. It would destroy all order if it could. Law must be inexorable. If men knew that they could depend upon anything but their own efforts, those efforts would not be made. We cannot expect, then, that God will alter his laws, or be turned from his purpose by man's prayers, and prayer can only be efficacious when, in accordance with mental law, it answers itself. Concentrated desire and earnest aspiration, attended by faith, has a strong tendency to answer itself. All power is will power, and at present we by no means know the limit to which our small borrowed portion of it can go. That it extends far beyond what is commonly supposed there can be no doubt, and fasting, within a certain limit, greatly increases this power, by withdrawing the force required for mental effort from vital processes.

More things are wrought by prayer
Than this world dreams.

—Alfred Tennyson.

The power of prayer, then, is very great, but it must be in accordance with law, and not in the breach of it, or in what is called "special providence." The noises that savages make with pots and pans and old kettles to avert an eclipse are not likely to have the desired effect; neither are the noises made to bring about rain or to alter any other of the laws of nature likely to produce any effect. But the noises that people make to bring about any mental or moral change are likely to be more efficacious; as, although not the most direct, they are themselves among the motives or causes by which the mind is moved. It is now pretty generally admitted "that all prayer for Divine intervention in, or modification of, events due to physical laws is absurd;" but this limitation to physical laws is only made in ignorance of the fact that the laws of the mental and moral world are as fixed and definite as the laws of gravitation and light, and we have no more right to expect God to interfere for our special benefit in one department than in the other. Throughout the universe there are fixed antecedents to fixed consequents, and it is much more religious to ascertain what these are, as they have been established by God, than to ask Him to break this order in our behalf. Beyond earnest desire and aspiration, and faith in the order of nature, prayer is a breach of that Reverence which alone is the proper attitude of true Religion.

It is upon our knowledge of, and obedience to, this fixed order, that our Health and Happiness depend. The Physical, the Moral, the Intellectual, the Social, the Political, the Economic laws—these are God's Will and Word; the extent to which they are revealed depending upon our own research, obedience to them constituting a sacred Religious duty. Veneration or Reverence has its natural expression towards parents and those who, when we are children, "are set in authority over us," as knowing more than we do; as children of a larger growth this feeling should be transferred to Natural Law, which is the automatic Will of the Greater Parent of all. "The further back we push the idea of a Creator, and the more we conceive His 'interference' to be limited to the ordaining of 'Laws,' the more certain it becomes that in these laws at least, if anywhere, we have the expression of his Mind and Will." * "I cannot too often repeat," says George Combe, "that unless the Christian morality be sustained and enforced by the order of nature, it is in vain to teach it as a rule of conduct in secular affairs. And how can this study be commenced and prosecuted, how can new truths be turned to practical account, except by reverencing Nature and her adaptations as Divine institutions, teaching them to the young, and enforcing them by the authority of the moral and religious sentiments."† In this ancient city of Coventry, containing some 40,000 inhabitants, the Sanitary Reformers, by the introduction of the Health of Towns Act, reduced the deaths from 27 in the 1,000 to 23; that is, they saved 4 times 40, or 160 lives annually, with about 18 cases of sickness that on the average attend each death. The labour was a hard one for the Sanitary Reformers, the Clergy and their adherents doing little to help, although they most assiduously visited the sick, fed the hungry orphans, gave spiritual consolation to the dying, and when dead buried them in "the sure and certain hope of a glorious resurrection." Theirs was the practice of true Christianity; the Sanitary Reformers represented the New Reformation—the practice of the Coming Religion.

Religion itself is the expression of simple Reverence and Trust, accompanied by Awe and Wonder, as we stand in the presence of constant, and unvarying, and irresistible Power; and all other phases of it are purely anthropomorphic, arising from the natural tendency we have to make God in our

* The Duke of Argyll, *Contemporary Review*, May, 1871.

† "Science and Religion," p, 178.

likeness, converting His Unity into a personality like our own.

Trust, faith that the future would be as the past, we could not have, unless we recognised God—that is, Intelligence and Plan in Creation. It has taken millions of years to bring the world to its present state of perfection, to produce all its varied life and enjoyment; and unless that were designed, unless there were both Intelligence and Power to support that purpose, there would be no reason that we can see why one turn of the screw—one additional change—should not make life again impossible in this world as in the beginning. All the *possible combinations* of our elements cannot have yet taken place, and one adverse combination might take us back to chaos. Let Nature but once be at fault in her arithmetic, and all life and enjoyment would again be at an end. And better thus, *infinitely*, than that *one* should continue to exist in endless suffering; that *absolute evil* should have been introduced into this world. Better blind chance, and the havoc it would make, than the life and immortality which Christianity, according to the orthodox creed, has brought to light, in the eternal enjoyment of the few and the endless misery of the many.

---

"'What is man capable of discovering and comprehending concerning God?'—is not a barren speculation, but one of a practical and important nature.

"Dr. Johnson defines the substantive '*Worship*' to mean 'Adoration; religious act of reverence;' 'to *worship*' is 'to adore; to honour or venerate with religious rites.' Again, 'to *adore*' is 'to worship with external homage.' Now, the external rites in which we embody our 'worship,' 'reverence,' or 'homage,' will obviously bear a relation to our motives in worshipping; and these will be influenced by our opinions of the character of the Being whom we adore. Tribes who ascribe the lower passions to their Deities institute immoral rites and ceremonies in honour of them. Those nations who regard God as cruel and revengeful, sacrifice animals and some of them men, to appease Him. Others, who ascribe to Him self-esteem and love of approbation, (their own predominant qualities,) offer him praise and glorification, and try to please him by expressing their own consciousness, (generally with much exaggeration,) of abject meanness and unworthiness.

"If I am right in saying that although God has not given us faculties fitted to comprehend Himself, yet He has given us powers which enable us *to understand His will in relation to ourselves and other beings over whom He has given us some degree of influence and control*, and that in the order of nature, He has revealed duties which we are capable of performing, then we may reasonably consider whether the rites of our religious worship should partake of the character of attempts to please God as a Being possessing human qualities, or be directed to do Him honour, reverence, and homage, by studying, expounding, and obeying His will as thus revealed to us. All existing forms of worship should be tried by their relation to what we can comprehend of the nature of God, and of His will. If without irreverence I might borrow an illustration from the relation between man and the lower animals, I should remark that it appears possible for one being to comprehend portions of the will of another, although he cannot conceive adequately the nature of that other. The dog, for instance, cannot comprehend the nature of the shepherd, but he can learn the shepherd's will to be, that he, the dog, should tend the sheep; and the dog, without attempting to know more of the shepherd's nature than this portion of his will, may obey it and preserve the flock. The horses which in our circuses are trained to dance, to fire pistols, to fetch tea-kettles, and to perform other surprising feats, do not comprehend the nature of the men who teach them to do these things, nor apparently do they understand the object or design of the actions themselves; but they seem to understand the will of the men, so far as it relates to the actions required of them, for they do the things they are taught. We should all agree that the dog sadly mistook his own capacities and his relations to man, if instead of hearkening to the shepherd's voice, obeying his will, and guarding the flock, he turned a deaf ear to one and set the other at defiance, and commenced a grand speculation on the *nature* of his master, and his *attributes*. We should be still more astonished at the want of a due sense of his own deficiencies and position, if the dog, in the midst of his speculation on this, to him, incomprehensible subject, and of his neglect of duty, ever and anon turned up his eyes and raised his fore-paws to his master, and uttered indications of intense admiration and veneration for him, calling him a being possessed of every faculty of canine consciousness in the highest state of perfection and in unlimited degree. And yet, ignorant and superstitious men do something analogous to this, when, instead of 'walking humbly' with God, studying His Institutions and obeying His will, they ascribe to him their own qualities, praise Him, and implore Him to protect them as His devoted worshippers; they all the while violating His laws. In the words of Dr. Fellowes ('The Religion of the Universe'), 'The only use which some religionists make of their understanding is to perplex it by inquiring into the nature of God. They leave the easy and feasible to attempt the impossible. They forsake the clear and simple to lose themselves in a region of clouds and darkness. For how can the finite hope to comprehend the infinite, the material the spiritual, the temporal the eternal? God can be known only in His works. THERE His agency is seen. THERE His will may be traced; there His laws be developed. But, what His nature is, or how He exists, must ever be past finding out. It is enough for us to know that He exists; but *how* He exists, it is vain, and indeed presumptuous to inquire.'"—("Science and Religion," by George Combe.)

## Section I.

### DEATH.

Like leaves on trees the race of man is found,
Now green in youth, now withering on the ground;
Another race the following spring supplies,
They fall successive and successive rise;
So generations in their course decay,
So flourish these when those have passed away.

We are such stuff as dreams are made of,
And our little life is rounded by a sleep.

---

"Thus I, considering everywhere
Her secret meaning in her deeds,
And finding that of fifty seeds
She often brings but one to bear," &c.

Ask what is death, and why? Are God and Nature, then, at strife! But first, can we answer the question, What is Life? Herbert Spencer says it is "the continuous adjustment of internal relations to external relations;" but these are only the conditions of its continued existence, and give no idea of what the "vital spark" is in itself. Schelling says, "Life is the tendency to individuation." I should say that Life is not only where the forces of nature are thus confined within definite limits, but also where they work towards a given object. But this again is only "its mode of action." But is not its mode of action all that we know of anything? Alas! we know not what Life is, whence it comes, or whither it goes. Whether, as Prof. Tyndall says, "not alone the more ignoble forms of animalcular or animal life, not alone the noble forms of the horse and lion, not alone the exquisite and wonderful mechanism of the human body, but that the human mind itself—emotion, intellect, will, and all their phenomena—were once latent in a fiery cloud;" or "whether, having waited until the proper conditions had set in, the fiat went forth, Let Life be," we do not know; but we do know that Life being here, Nature has made wonderful provision that the spark should not be blown out. An organic being is the result of all the forces and conditions which have been transmitted from frame to frame, each an improvement on the other, and nature has most bountifully provided that such improvements, the result of so much time and care, shall not be lost, but carried on. She brings forth at least fifty seeds for every one she is able to rear, and Life is kept at so high a pressure that there is not a plant or animal whose produce if left unchecked would not of itself soon cover the earth. "There is no exception to the rule," says Darwin, "that every organic being naturally increases at so high a rate, that if not destroyed, this earth would soon be covered with the progeny of a single pair." The rate of increase is geometrical. We have population always pressing on the means of subsistence, and this has been the mainspring of all progress and all order; for the most perfect order and arrangement exist in the mode in which life is kept within due bounds: the good and strong preserved, the weak destroyed, the object evidently being to keep the largest possible number in the greatest possible strength and vigour and capability of enjoyment—no respect being paid to our little individualities. In the scale of being, by a most systematic provision, we are all made *comfortably** to fit into each other. The conversion of what is lower into what is higher is always

* Perhaps it may be thought that "comfortably" applies as little in this case as when the hangman said his gallows would hold three comfortably; but the pain of death has always been greatly exaggerated for theological purposes.

going on, each animal is eaten and eater—end and beginning in succession. "The perch swallows the grub-worm, the pickerel swallows the perch, and the fisherman swallows the pickerel; and so all the chinks in the scale of being are filled." Man in his wisdom often tries to break this chain. He exterminates the crow, and the grubs eat his crops; he kills the little birds, and his crops fail altogether; he kills the owl, and mice abound; he kills the weasel, and he is eaten up with rats; and although the connection is less direct and the effects of less importance, if he were to get rid of his old maids, he would have no red clover or heart's-ease, for the old maids keep cats, and cats kill the mice, and the mice, if allowed to live, destroy the humble-bees, who alone are able to fructify and spread the red clover and heart's-ease. Near villages and small towns the nests of humble-bees are found to be more numerous than elsewhere, owing to the number of cats which destroy the mice. Alas for the butterfly, our emblem of immortality! however much he may dream of the life to come, he lives only till a new generation is begotten. His immortality is very short; but as he lives on nectar and love, he may consider that better than an eternity as a mere grub. His wife has so many children that it would destroy all man's hopes of "cabbage" if it were not for the ichneumon fly, who kills the caterpillars as they are hatched from the eggs. The oak alone is estimated to feed at least 200 kinds of caterpillars, so that the butterfly is immortal in the race, if not in the individual. Death is thus the parent of life: something always dies that another may live; one superabundant population supporting the life of other description of beings. But the new edition of life has generally been greatly improved; "the structure of an organism being a product of an almost infinite series of actions and reactions, to which all ancestral organisms have been exposed."* This conservative power of greatly improved transmitted influences has

* Herbert Spencer.

again to contend with its surrounding conditions, which ultimately again prevail, but not before all worth preserving has been passed into new bodies: every vacant place being filled up by some creature capable of increased enjoyment. "I disbelieve in the destruction of anything," says the author of the "Pilgrim and the Shrine," "of even a thought—but leaves a germ, idea, or cell of greater capacity than had before been possible, as a worthy result of the whole previous universe of being." "All death is nature in birth—the assumption of a new garment, to replace the old vesture which humanity has laid aside in its progress to higher being."*

In this way is Death the parent of Life, and this is necessarily so, for Life is Force, and obeys all the laws of force, and the "invariability of the sum of all the energies of the universe forms the doctrine known as the Conservation of Energy." Thus Force is a fixed quantity, so that if it begins in one place it must necessarily be taken from another, or cause death elsewhere. "Each portion of mechanical or other energy which an organism exerts, implies the transformation of as much organic matter as contained this energy in a latent state," and "this organic matter yielding up its latent energy, loses its value for the purposes of life, and becomes waste matter needing to be excreted."† It is dead, but it has been the parent of new life. All work implies waste; that is, organic matter yields up its latent energy, is exhausted and dies. "We cannot live, as to our total organism, unless we are always dying as to our atoms; * * every thought involves the death of some particle of the brain;"‡ or, as Huxley says, "so much eloquence, so much of the body resolved into carbonic acid, water, and urea." This waste may be repaired, but the power of repairing is impaired with every exercise of

* "Memoir of Fichte," by W. Smith.

† "Principles of Biology," vol. 1, p. 176, by Herbert Spencer.

‡ Grindon, p. 34.

it. Thus old age and death. But "where energy disappears," says Prof. Tyndall, "it changes its form merely; and if we are intelligent enough to follow it, we shall find it somewhere, unaltered in amount, though perhaps with an entirely new face."

This, then, is Death. Life proceeds only from Life. A part of some already living organism is thrown off as a seed or germ, "the structure of such organism being the product of an almost infinite series of actions and reactions to which all ancestral organisms have been exposed." This "individuation" of special powers derives its force or life from other organic matter yielding up its latent energy—that is, from the death of something else. This "individuation," or organism, is merely an instrument, or machine, which is supplied with energy in one form, and converts it into another according to the structure or law of the machine. The machine wears, and the power to repair this waste gets less and less with the constant use of the machine, until at last the "tendency to individuation" is overcome, and the specialised forces are again added to the other forces of nature from which they were originally taken or derived.

We have thus a network of nerve spread over the whole world, specialised into organisms of varying powers and functions. Yet, as Professor Huxley tells us, a unity of composition pervades the whole living world; the difference between the lowest in the scale, both in plants and animals, being difference in degree, not of kind. All are subject to the same laws of birth, growth, decline, and death, and "at the earliest stages of its development no human power can distinguish the human ovum from that of a quadruped."*

These countless millions of nerve cells are constantly supplied with fresh force by the life of something else given up to them, and if the old and worn-out forms are not quickly removed the new ones could not exist. This is precisely the

* "The Physiology and Pathology of Mind," p. 56, by H. Maudsley.

same whether it relates to a cell or to an aggregate of countless thousands of cells forming an organism. A body is composed of cells, the great network of nerves; the great body of sensibility consists of individuals composed of an aggregate of cells; but the laws of waste and supply, of life and death, are precisely the same in each. Thus death in the aggregate is simply a change—a correlation of force, by which more vigorous life is secured. In fact, there is no such thing as death. Individual cells pass in and out of the body; individual bodies pass in and out of the great organism for the production of sensibility, but not death, but the life and sensibility are more and more; the process being one merely for the conversion of what we call physical force into mind or consciousness. "All death in nature, is life, and in death appears visibly the advancement of life. It is not death which kills, but the higher life, which concealed behind the other, begins to develop itself. Death and birth are but the struggle of life itself to attain a higher form."* "There is no death in the concrete; what passes away passes away into its own self—only the passing away passes away."† Dr. Henry Maudsley says: "The nerve-cell of the brain, it might in fact be said, represents statical thought, while thought represents dynamical nerve-cell, or, more properly, the energy of nerve-cell."‡ And again: "Without doubt the Will is the highest force in Nature, the last consummate blossom of all her marvellous efforts." * * Thus the force of the great men who worked in harmony with the current of events among which they lived "was a force not their own; the power of the universe was behind them, and they became the organs of its manifestation. * * If we ask whence the impulse that displays itself in this upward *nisus*, we can only answer lamely that it comes from the same unfathomable source as the impulse that inspires or moves

* Fichte's "Destination of Man," p. 127. † Hegel.
‡ "The Physiology and Pathology of Mind," p. 45.

organic growth throughout nature." In fact, animal bodies are but the organs through which Universal Mind acts: they are the machines for the conversion of mind, which has become automatic, again into Consciousness and Will. They

> Are but organic harps divinely fram'd,
> That tremble into thought, as o'er them sweeps,
> Plastic and vast, one intellectual breeze,
> At once the soul of each, and God of all.
> —Coleridge.

But Death to man is the extinction of his Individuality, of his personal Identity, of the Ego which is the centre of the universe to him, around which everything else revolves. It is evident that if man had made himself he would have made himself immortal; but he would probably have made a great mistake, as he always does when he thinks himself wiser than his Maker. What He has ordained is a succession of Beings, as opposed to the immortality of one. Humanity is immortal, not man; and in the great Body of Humanity, with its Soul, the principle of Sensation, all is preserved that is worth preserving: while all dies with man that is in the way of progress, all vice and error are at once reformed, all prejudice dispelled,* and "in the mortality of man lies the salvation of truth." Everything dies, from a world

* During the advance from maturity to old age "the habits of life grow more and more fixed; the character becomes less capable of change; the quantity of knowledge previously acquired ceases to have its limits alterable by additions; and the opinions upon every point admit of no modification." Societies pass through the same successive phases.—("First Principles," pp. 187, 188, Herbert Spencer.)

"By the necessity of the case almost, an old man becomes conservative, and the *laudator temporis acti;* for the evolution of events goes on when his nature has ceased to assimilate and develop; he has accordingly no sympathy with them; but, retreating within the shell of a calcified past, obstinately brands as revolutionary what is truly *evolutionary.*"—("The Physiology and Pathology of Mind," p. 219, Dr. H. Maudsley.)

to an atom, from a man to a monad, that it may be born again in increased beauty and in increased capability of enjoyment. Our world was probably formed out of nebulous matter; and we can see evidence that it must get old and die like everything else. Its stores of force in its coal strata are being used up; its mountains, upon which its circulatory system and vitality depend, are gradually crumbling before the action of the elements; the internal heat is lessening, the centripital force is increasing, and it must ultimately give up its individuality by becoming a part of a larger system in the sun, or be dissipated by the collision and arrested motion into the nebulous matter and fiery mist in which it first began.*

* "In a very remarkable, but seemingly little known treatise, Kant expounds a complete cosmogony, in the shape of a theory of the causes which have led to the development of the universe from diffused atoms of matter, endowed with the simple attractive and repulsive forces.

"'Give me matter,' says Kant, 'and I will build the world;' and he proceeds to deduce from the simple data from which he starts, a doctrine in all essential respects similar to the well-known 'Nebular Hypothesis' of Laplace. He accounts for the relation of the masses, and the densities of the planets to their distances from the sun, for the eccentricities of their orbits, for their rotations, for their satellites, for the general agreement in the direction of rotation among the celestial bodies, for Saturn's ring, and for the zodiacal light. He finds in each system of worlds indications that the attractive force of the central mass will eventually destroy its organisation, by concentrating upon itself the matter of the whole system; but, as the result of this concentration, he argues for the development of an amount of heat, which will dissipate the mass once more into a molecular chaos, such as that on which it began.

Kant pictures to himself the universe as once an infinite expansion of formless and diffused matter. At one point of this he supposes a single centre of attraction set up, and by strict deductions from admitted dynamical principles, shows how this must result in the development of a prodigious central body, surrounded by systems of solar and planetary worlds in all stages of development. In vivid language he depicts the great world-maelstrom widening the margin

We have a similar process to what we call death, on a smaller scale, always before us. By a silent and imperceptible force the water of the sea is raised in vapour, then descends in clouds upon our mountains and upon our lands, forming the circulatory system of the world, and giving life to plants and animals, and then, after doing innumerable other work flows on to the great ocean, again to commence its vitalising course. The river is immortal, for it is the same river flows on for ever, although not one drop is the same; but had it been immortal in man's sense of being *the same*, instead of a succession, we should have had stagnation, instead of the clear, bright, living water. So it has been held by Buddha, and the greatest among the Hebrew and Greek minds, that God is the great ocean from which all else proceeds—that the world is a manifestation of God, emanating from him, and being re-absorbed into him again. "As a drop of water pursues its devious course in the clouds, in the rain, in the river, a part of a plant, or a part of an animal, but sooner or later invariably finds its way back to the sea from which

of its prodigious eddy, in the slow progress of millions of ages, gradually reclaiming more and more of the molecular waste, and converting chaos into cosmos. But what is gained in the margin is lost in the centre; the attractions of the central systems bring their constituents together, which then, by the heat evolved, are converted once more into molecular chaos. Thus the worlds that are, lie between the worlds that have been, and the chaotic materials of the worlds that shall be; and in spite of all waste and destruction Cosmos is extending his borders at the expense of Chaos."—(From Huxley's Address to the Geological Society, 1869.)

"The doctrine of heat brings out more clearly than any previous scientific theory, that whereas all life, all play of phenomena, consists in interchange, the inevitable course of nature is towards dead uniformity by the equal diffusion of heat, the destruction of difference of motion by friction and collision, and the gradual establishment of the most stable chemical compounds. The actual cosmos must, it seems, have started supernaturally at a definite past time, and will, if allowed, run out its series of changes at an incalculably remote but definite date."—(*Contemporary Review*, May, 1868, p. 68.)

it came, so the soul, however various its fortunes may have been, sinks back at last into the Divinity from which it emanated."* Then we have the life of vegetation born in fresh green loveliness every Spring to die in Autumn and prepare a soil for re-appearance in increased beauty with the new year; and man has his Spring, Summer, Autumn, and Winter, with the varied happiness that attends each season, and the increasing fertility of mind as each generation succeeds another. But man thinks this a great mistake, or rather a great "fall." He would have made himself immortal. He would have had one everlasting Spring or Autumn, with the comparative stagnation that must necessarily attend it. And for what? Why, that Dick Snooks, being in possession, might continue to live instead of Tom Styles, and thus unfairly monopolise all the enjoyment to himself. We have the swallows every year, and the lark, with his song as sweet as it was a thousand years ago, and who thinks of inquiring whether they are the same swallows and the same lark as the year before? There are the same song and the same enjoyment, and that is all that is worth continuing, and what does it matter, therefore, in what bodies they appear. Nature's plan is to be careful of the type, but the bodies waste with work, deteriorate with age, and they are all ground up periodically to begin life again in some of its infinitely varied forms. † There is nothing

* Dr. J. W. Draper.

† Perhaps among the quickest instances of this re-conversion is the wonderful one mentioned by the Rev. Canon Kingsley, in his "Letters from the Tropics," *Good Words*, June, 1870. He says:—"Under the genial rain and genial heat (Trinidad) the timber tree itself, with all its tangled ruin of leaves and parasites, and the boughs, and stems, and leaves snapped off, not only by the blow, but by the very wind of the falling tree—all melt away swiftly and peaceably in a few months—say almost in a few days—into the water, and carbonic acid, and sunlight, out of which they were created at first, to be absorbed instantly by the green leaves around, and, transmitted into fresh forms of beauty, leave not a wrack behind."

now living, perhaps, that has not lived in some form or other at least a hundred times before. The organism melts into air, or descends into the earth, again to ascend, through plants, and form the living system of horse, ox, or man; or it is devoured by other beings, and commences through them a higher life at once. As to man, some forty millions of human beings come into existence every year, and some thirty millions pass out, and the great body of Humanity is thus ever on the increase in size and capability of enjoyment, and nothing is lost by this change worth retaining. No doubt every man has lived at least a dozen times before on this earth, and nothing has been lost to him but the sense of this previous existence: perhaps even that has not been all lost;—

Our birth is but a sleep and a forgetting:
The soul that rises with us, our life's star,
Hath had elsewhere its setting,
And cometh from afar;
Not in entire forgetfulness,
And not in utter nakedness,
But trailing clouds of glory do we come
From God, who is our home.

If we calculate man's existence upon this earth at 6,000 instead of the 100,000 years now recognised by geologists, it has been calculated that the earth would be covered with graves, each grave containing 128 persons.† But it is from

† It has been calculated by good arithmeticians that the number of persons who have existed on our globe since the beginning of time amounts to 36,627,843,273,074,256. Rather a large number to tell us all their thoughts as well as actions on the Day of Judgment. These figures, when divided by 3,095,000, the number of square leagues on the globe, leave 11,320,680,732 square miles of land; which being divided as before, give 1,314,622,076 persons to each square mile. If we reduce these miles to square rods, the number will be 1,853,174,600,000, which divided in like manner, will give 1,283 inhabitants to each square rod; and these being reduced to feet, will give about 5 persons to each square foot of *terra firma*. In this vast cemetery, on each

the air that plants derive their principal sustenance, and animals feed on plants, and every particle of the atmosphere, therefore, must once have been living man. "To fancy, as many do, that death is not only terrible and affrighting, but physically painful, is quite a mistake, being to look for sensibility in the *loss* of sensibility. Death is a sleep rather than a sensation; a suspension of our faculties rather than a conflict with them: instead of a time of suffering, a time of deepening unconsciouness. Dr. Baillie tell us that his observation of death-beds inclines him to the firm belief that 'nature intended we should go out of the world as unconsciously as we came into it.'"*

It is thus evident, that as far as our own observation extends, Nature's "plan" is best accomplished by a succession of beings rather than by the continued existence of one. The "individuation" upon which life depends is transmitted from one body to another in organic structure upon which its *specific* powers of acting depends, the forces, both bodily and mental, which it displays being all derived from the elements around us. This is the "order of Nature" in this world, but man thinks that in some far distant world, to which at death he will be conveyed, all this order will be reversed in favour of his retaining his Identity. He believes that the speciality of the body may be destroyed, but that the speciality of

square rod of it, 1,283 human beings lie buried, each rod of it being scarcely sufficient for 10 graves, with each grave containing 128 persons. The poet says—and taking the estimate at 100,000 years instead of 6,000, and supposing the atoms to have been used but once, it is not mere poetry:—

> There 's not one atom of yon earth
> But once was living man;
> Nor the minutest drop of rain,
> That hangeth in its thinnest cloud,
> But flowed in human veins.

* Grindon, p. 278.

the thinking principle, or Soul, is indestructible. Of course the Soul, like everything else, is immortal; the question, however, is, does its speciality, like that of the body, depend upon the organism with which it is temporarily united? The poet says:

> Whether the Sensitive Plant, or that
> Which within its boughs like a spirit sat,
> Ere its outward form had known decay,
> Now felt this change, I cannot say.
>
> Whether that lady's gentle mind,
> No longer with the form combined
> Which scattered love, as stars do light,
> Found sadness, where it left delight,
>
> I cannot guess; but in this life
> Of error, ignorance and strife,
> Where nothing is, but all things seem,
> And we the shadows of the dream,
>
> It is a modest creed, and yet
> Pleasant, if one considers it,
> To own that death itself must be,
> Like all the rest, a mockery.
>
> That garden sweet, that lady fair,
> And all sweet shapes and odours there,
> In truth have never passed away:
> 'Tis we, 'tis ours, are changed! not they.
>
> For love, and beauty, and delight,
> There is no death nor change; their might
> Exceeds our organs, which endure
> No light, being themselves obscure.
>
> —Shelley.

On the question of the Immortality of the Soul—of the retention of our Identity and Personality in another world; of whether the Soul be the result of organisation, and dependent for its duration on the duration of conditions, or whether it be in itself capable of self-existence, I leave every man to his opinion. The theory of an individual separate entity, distinct from the body, and differing in essence, is attended with many difficulties. If they really differ in essence how can one act upon the other, that is, matter upon that which is immaterial, &c.? But we may leave that, as all now admit that we know nothing of essences, and therefore nothing of any intrinsic difference between body and soul; but the question is, when and where, and from whence, does the "soul" originate? Darwin says: "In a series of forms graduating insensibly from some ape-like creature to man as he now exists, it would be impossible to fix on any definite point when the term 'man' ought to be used,"* and until that is settled, I presume we cannot "fix on any definite point" when man gets his soul. Darwin also says, speaking of the extinction of race: "When one of two adjoining tribes becomes more numerous and powerful than the other, the contest is soon settled by war, slaughter, cannibalism, slavery, and absorption. Even when a weaker tribe is not thus abruptly swept away, if it once begins to decrease, it generally goes on decreasing until it is extinct." † Now, if these interesting creatures are men, I presume they have souls; but is the difference between them and civilised men dependent upon their souls or bodies? It is not clear of what use the souls are; it is quite clear it is not for their advancement, as the tribe becomes extinct. Sir J. Lubbock remarks how improbable it is that our earliest ancestors could have counted as high as ten, considering that so many races now in existence cannot get beyond four. Now, these souls which cannot count four, can scarcely have the dignity usually assigned to them; they must be an inferior sort of souls, as a magpie can count nine. The same difficulty presents itself in the fœtus—at what particular period of gestation does the fœtus get its "living soul?" The Roman Catholics, in certain cases, think it necessary to

* "Descent of Man," vol. 1, p. 235.

† Ibid. p. 238.

baptise the infant *before* it is born, lest the devil should afterwards lay claim to its soul. The Mormons have not, I believe, settled this point; but they hold that there are always a stock of souls on hand, waiting to be born, and that the highest religious rite and duty is to beget tabernacles of the flesh for these immortal spirits waiting to be born.*

Augustus de Morgan tells us, in those humorous Paradoxes of his *(Athenæum)*, that "it has been started that there is, somewhere or another, a world of souls which communicate with their bodies by wondrous filaments of a nature neither mental nor material, but of a *tertium quid* fit to be a go-between; as it were a corporispiritual copper encased in a spiritucorporeal gutta-percha. My theory is that every soul is everywhere *in posse*, as the schoolmen said, but not anywhere *in actu*, except where it finds one of its bodies."

Hegel avoided any definite expression of opinion on the Immortality of the Soul. He recognised in the Life of the Universe only God's self-development, and in the life of individuals only the ever-changing and evanescent "forms" and "modes" of that evolution. From the whole spirit of his writings, we gather that he did not believe in the continued existence of the individual; and Freidrich Richter says that Hegel's principles give no countenance to what he calls this "ambitious craving of egoism." Strauss and Michelet appear to hold the same view, and Prof. Owen says, "philosophy does not recognise an immortal entity, mental principle, or soul." Life is the "vital spark of heavenly flame," which is the great mystery—at present we know not whence it comes or whither it goes. It belongs to plants and animals. No doubt the Universal Life and Universal Soul are the same.

But whatever opinion we may form, it is at least certain

* "New America," p. 288, by Hepworth Dixon.

that "where we are, Death is not; and that where Death is, we are not;" and that the worst that can happen, should there be no "future state" for us, will be that we shall return to what we were before we were born, when, as far as I recollect, we did not much regret that we were not alive; and this would be infinitely better than that any one, much more the majority of mankind, should continue to exist in endless suffering, according to the popular creed. However vivid and pleasant the recollection in another world of our previous existence in the grub-state here, it would not be worth this endless woe to others. I have infinite faith in God's Providence, and that the immense amount of enjoyment He has created will be retained and increased, whether in this world or any other; and for myself, as an individual, all I can say is, "If it is better, *all things considered*, that I should continue to exist in some other world, I believe I shall be certain to do so; if, all things considered, it is not better, I do not wish the order of God's Providence, by which the largest amount of enjoyment is produced, to be changed for me. Before I was born I did not crave for existence, and I cannot do so after I am dead: I cannot even envy the person, whoever he may be, or whatever he may be called, who is then taking *his turn* in Life: for let us not forget that without death all creatures in the world at present could never have lived — the space would have been filled up. Death to others has been life to us, and our death will be life to others. We have had our time, and

> Not heaven itself upon the past has power.
> And what has been has been, and I have had my hour.
>
> —Dryden.

No; if it is God's Will,

> I to the elements
> Resign the principles of life they lent me.
>
> —Sir Walter Scott

And with the aid of Jean Paul Richter, I here write my Epitaph:

"Great Nature, to Thee thou lettest me always come when among men troubles gathered round me! Thou wert my oldest friend and my truest, and thou could'st always console me, until I fell from thy arms at thy feet, and needed consolation no more."

"The weary are at rest."—(Job.)

PERMITTE CŒTERA DEO.

## CHAPTER VI.

### SOCIOLOGY.

It is the function of Moral Science to frame the laws by which men can live most happily together; and of Social Science to establish the Institutions or conditions in which those laws can be best put into practice or administered.

All animals have a natural tendency to increase faster than their means of subsistence, but as a rule they do not starve to death, as they live on one another. Man is no exception to this law. It is only, however, among savages that men feed on each other. Civilised men eat up each other's means of subsistence, so that many necessarily die of starvation, or if not directly of famine, of the neglect and disease and ignorance which poverty always engenders. Half the children born die before they are five years of age, and the average term of human existence is about thirty-three years, instead of, as it should be, twice that time. Malthus showed beyond all question that man possesses a power of increase beyond the diminution caused by death, and that this increase was greater than his food could be made to increase on the same area. Consequently man must either limit this natural power of increase by moral or prudential checks, or he must constantly be taking fresh land into cultivation. Failing these, he must die of starvation and the deteriorating influences engendered by it. At present he does all three in about equal proportions. Nevertheless, as I have previously pointed out, from the beginning this pressure of population on the means of subsistence has been the great mainspring of all progress. It has been the struggle for life itself that has forced into activity every intellectual faculty, and has spread mankind over all the globe to the cultivation of new, wild, and barren regions, when they would have much preferred staying at home. It is this law, however hidden and obscured, that has framed all human institutions, Social, Moral, and Political.

As the world is not yet occupied and brought into full cultivation, the means of subsistence never need fall short of even the natural rate of increase of the population if proper arrangements were made for production and distribution. Machinery now does an immense amount of work, and if the produce of a manufacturing country can be freely exchanged for the produce of land in another, all may be well off. But the people in possession, who hold the land and machinery, will not allow those who come into the world without either, to make use of their property except upon their own terms. Hence has arisen the great inequality in the distribution of wealth. If some workmen are sufficiently well off to be able to refuse to take less for their labour than a fair share of what they are instrumental in producing, the natural rate of increase of the population will always furnish some one who is willing—in fact who is obliged to take whatever is offered to him or starve. It is this competition among workmen themselves that has hitherto enabled the capitalist to dictate his own terms and to regulate the rate of wages. The position admits of very simple illustration. Suppose two men with potato plots, and one of them had a spade (machinery), which he offered to lend to the other man on condition that he should dig and plant both lots. It would evidently be the interest of the latter to accept these conditions; and although he did *all* the work he would get at least half the produce, and that upon easier terms than if he had to set his potatoes without a spade. But suppose the man with the spade was the stronger, and sought a cause of quarrel with the other, and turned him out of his allotment. Having then no other means of subsistence, the dispossessed man

There has been very little agreement among political economists as to what ought to constitute the labourers' share of joint produce. Adam Smith says, "the *produce* of labour constitutes the natural recompense or wages of labour;" Ricardo, that "that which is sufficient to place the workmen, one with the other, in a position to exist, and to propagate his species, without increase or diminution, is the true natural measure of the natural rate of wages; while Herr von Thüner tells us that "the labourer is valued at the cost of his bringing up, as the machine is at the cost of its construction, and the beast of burden at that of its rearing." Lord Derby, when Lord Stanley, took a much more generous and right view of the question; he told the Glasgow students that "he thought a scrupulous and high-minded man would always feel that to pass out of the world in the world's debt—to have consumed much and produced nothing—to have dined, as it were, at the feast, and gone away without paying the reckoning—was not, to put it in its mildest way, a satisfactory transaction, however unimpeachable it might be, and rightly so, in the eyes of economic or social law." A statesman's labour is worth a great deal, still it may be fairly questioned whether Lord Derby's labour is equal to the revenues of his House—his production equal to his consumption. It is true he does not himself consume all the large share allotted to him, that he must share it with many others; still it is for the most part consumed non-productively, or leads to the production of many useless things, instead of the necessaries the community require. Mr. Mill holds that the rise in the value of land, which results from the general growth in the community in wealth and population, properly belongs to the state, the quantity of land being limited, and its possession therefore a monopoly. "This proposition," says the *Pall Mall Gazette* (May 31, 1871), "looks rational enough and harmless enough, but it is impossible to believe that it can be stated by such an authority without greatly strengthening the 'labour reformer's' doctrine that nobody should pocket any profit, rent or interest, which is not the product of his personal labour, which, laughable as it sounds, is devoutly held by an increasing number of English and Continental workmen, and which, although the seed of a barbarism compared to which that of the dark ages was respectable and hopeful, is yet destined before it disappears to give the civilisation of the Western world a severe shaking."

At the present time Machinery has greatly increased Capital or Production, and a greater approach to Freedom of Trade has facilitated Distribution, and this has greatly improved the condition of the labourer; still competition keeps down his rate of wages, and he nowhere receives a fair share of the joint produce: the system still continues to create enormous wealth on one side and comparative poverty on the other. It is evident that as steam and machinery do most of the work, the owner of the machine, upon the present system, gets the largest amount of the produce of work. Prof. Jacoby, speaking to his constituents at Berlin of this tendency of capital to accumulate in a few hands, and to leave at least a part of the working proletariat exposed to great distress, says:

"In such a condition of things, it is the incumbent duty of every honest and thoughtful man to put to himself the following question:

"How can we modify the present relations of society and property so as to realise a more equitable distribution of the common revenue, and to obviate the distress of the working classes, which daily assumes more extended proportions?

"How can we, without restricting the liberty of labour, and without prejudicing the progress obtained by industry (on a large scale), realise a more equitable distribution of the common revenue, and one more suited to the interest of *all*?

"The answer for us at least cannot be a doubtful one; there is but one means which can lead to this end: *The abolition of the wage system and the substitution in its place of co-operative labour.*

"Whoever can read the signs of the times will not deny that this is the thought which more or less consciously is at the bottom of all working men's movements in every country of Europe. Just as slavery and serfdom, which were also formerly held to be *necessary* social institutions, have everywhere given way to the wage system, so impends to-day a revolution of the same kind, and not less important—namely, the transition of the wage system to labour, and labour free and equal in the right of association. It is needful so to act that this revolution be effected in the most peaceful manner, which cannot happen except by the unanimous concurrence of all the social forces interested in it. * * * * * *

"Let us now briefly summarise what we have said:

"The wage system answers now as little to the exigencies of justice and humanity, as slavery and serfdom in former times.

"Just as it was with slavery and serfdom, the wage system was formerly a progress by which society has derived incontestable advantages.

"The social question of our times consists therefore in the abolition of the wage system, without prejudice to the advantages resulting from the common labour of great collective industry.

"There is for this but one means, the system of free labour by association—the co-operative system. The present time is a transition period from the wage system (system of production by means of capital) to the system of labour by association.

"In order that this transition may be effected in a peaceful manner, it is requisite that the workmen, employers, and the State act in common.

"It is the duty of workmen to unite, in order to resist the oppression of capital and to raise themselves by education to moral and material independence.

"It is the duty of the employer to engage himself in the cause of the workman's well-being in a philanthropic spirit, and especially to accord to him a share of the profits of labour.

"Finally the State, by the protection of association, by fixing the hours of labour, and by giving gratuitous instruction, ought to favour the efforts of workmen towards civilisation. Upon the State devolves, at the same time, the duty of protecting the system of production by association on a large scale, of a reform in the system of banks of credit, and of the institution of State Credit?

"As such help can only be expected from a free State, it is clear that the workmen and their friends must, before all, procure for themselves political liberty.

"Political liberty, social liberty, liberty of the citizen, without sacrificing the majority of mankind as wage-labourers—this is the problem of our era.*

Thomas Carlyle has long been of Prof. Jacoby's opinion; he says:

"All human interests, combined human endeavours, and social growths in this world, have at a certain stage of their development required organisation; and work, the grandest of human interests, does now require it."

"This that they call 'Organisation of Labour,' is, if well understood, the problem of the whole future, for all who would in future pretend to govern men."

"We find," he says, "all mankind heaped and huddled together, with nothing but a little carpentry or masonry between them; crammed in like salt fish in their barrel;—or weltering (shall I say?) like an Egyptian pitcher of tamed vipers, each striving to get his head above the rest."

"What wonderful accessions have been made, and are still making, to the physical power of mankind; how much better fed, clothed, lodged, and, in all outward respects, accommodated men now are, or might be, by a given quantity of labour, is a grateful reflection which forces itself upon every one. What changes, too, this additional power is introducing into the social system; how wealth has more and more increased, and at the same time gathered itself more and more into masses, strangely altering the old relations, and increasing the distance between the rich and poor, will be a question for political economists, and a much more complex and important one than any they have yet engaged with."

The present system of "unlimited competiton," or every one for himself, and the devil take the hindmost, gives us a million paupers who are fed at the public cost,† and to these

* A Speech delivered by Deputy Johann Jacoby to his Constituents of the Second Arondissement of Berlin, on the 20th January, 1870. Printed for private circulation.

† The Poor Rate has increased, during the last ten years, 29 per cent in Ireland, 32 per cent in England, and 34 per cent in Scotland. The total expenditure for poor relief in the United Kingdom is £9,793,000, or 6s. 4d. per head of the population. The rate per head of the population is exactly double in England to what it is in *poor* Ireland.—(*Pall Mall Gazette*, August 19, 1871.)

may probably be added a million criminals who feed *themselves* at the public cost. Quetelet, however, truly says: "Society prepares crime, and the guilty are only the instruments by which it is executed."

Our imperious law of Political Economy, Supply and Demand, never pays a man according to what he produces; and a brass button or a ribbon going out of fashion reduces whole towns and communities to poverty and destitution. As luxury increases this uncertainty increases also. "We all know the periodical distresses of our staple industries," says Dudley Baxter. "It may be said that these were accidents. They are not mere accidents, but incidents, natural incidents, of our manufacturing economy. They are sure to recur under different forms." Over-production, over-speculation, change of fashion, foreign competition, are constantly putting our workmen's means of living in danger; and the great increase of pauperism and crime, and the numbers at present out of employ, consequent on the panic and commercial disasters of 1866, are inducing some of our leading journals to doubt whether the *laissez-faire* doctrine in its extreme application is the right thing after all. At a large meeting held at Exeter Hall on the 12th April, 1869, for the purpose of promoting emigration, Mr. Torrens, M.P. for Cambridge, declared that not hundreds or thousands but millions must be taken off the surplus labour of the country before adequate relief can be afforded. In other words, says the *Pall Mall Gazette*, he desires that the government should do for England what famine did for Ireland, that is, deliberately take a large proportion of the poorest classes and transplant them to Canada or Australia. "If this be sound," adds the same journal, "it is certainly a most humiliating confession; it amounts to saying that the growth of commercial prosperity in the last half century has had for its nett result the production of an unparalleled mass of misery alongside of great wealth in individual cases—a misery, indeed, so great that nothing short of a heroic remedy will be sufficient." The truth seems to be that our economic system has divorced one-third of the population of Great Britain from their food, and the land upon which it is grown, and it is highly probable that we shall have to bring them together again. The same system produces the same results elsewhere. In December last Mr. David A. Wells, Special Commissioner of Revenue in the United States, presented a report to Congress, the general purpose of which was to prove that "the rich were growing richer and the poor poorer," and Mr. Horace Greely estimated last January that there were half a million persons within sight of the city steeples of New York who were hungry, destitute, and out of work. This evidence is confirmed by the report of the secretary of the New York Association for improving the condition of the poor, and by a number of manufacturers and other employers. The results of the manufacturing system are thus described in a Senate document, No. 44, January, 1869:—"A helpless crowd of workers, the oppression of low wages, inevitable poverty, and a disguised serfdom, a rich master, a poor servant, and a mean population. Such is the story of manufacturing in Old England, and such is the story of manufacturing in New England." Mr. Wells finally shows that the conclusions drawn from savings-banks and insurance statistics rather support than refute the statements which he has laid before the public. He does not discuss the condition of the West, but expresses his belief that "like causes have produced like effects" everywhere.

Mr. Dudley Baxter, in summing up his admirable and exhaustive work on our National Income, says: "The Income of England is the largest of any nation, and shows wonderful good fortune and prosperity; but we must not forget that it rests upon an unstable foundation. The turn of trade, or obstinacy and short-sightedness of our Working Classes, or a great naval war, may drive us from the markets of the world, and bring down our Auxiliary as well as our Productive industries. In our present complex civilisation the effect of such a calamity on a large scale can hardly be imagined. We might see our national income disappear far more rapidly than it has increased, and a period of suffering among our population of which no cotton famine or East London distress can afford an adequate idea. England's position is not that of a great landed proprietor, with an assured revenue, and only subject to an occasional loss of crops or hostile depredations. It is that of a great merchant, who by immense skill and capital has gained the front rank, and developed an enormous commerce, but has to support an ever-increasing host of dependents. He has to encounter the risk of trade, and to face jealous rivals, and can only depend on continued good fortune, and the help of God, to maintain himself and his successors in the foremost place among the nations of this world."

Mr. Henry Fawcett, M.P., in his late work, "Pauperism; its Causes and Remedies," gives his testimony also very much in the same strain. He says:

"Unless the generally recognised principles of economic science are a tissue of fallacies, it can be easily shown that no scheme of social improvement can be of permanent efficacy if it is unaccompanied by an increased development of providence amongst the general mass of the people. The truth of this is proved in a very striking manner by considering the effects which have resulted from free trade. If any one refers to the speeches which were made during the Anti-Corn-Law agitation by the prominent advocates of the movement, he will find that the most glowing anticipations were indulged in with regard to the consequences which would result from the abolition of protection. Mr. W. J. Fox, who was perhaps at one time the most distinguished orator of the party, when addressing a large meeting in Covent Garden Theatre, asserted that the abolition of protection would exterminate pauperism; and he predicted that in a few years the ruins of the workhouses would mark the extinction of protection just in the same way as the ruins of the baronial castles mark the destruction of feudalism.

"Nearly a quarter of a century has now elapsed since the adoption of free trade in corn, and the present state of London, with its 160,000 paupers, is a cruel comment upon these prophecies. But how is it that what so many confidently expected has so entirely failed to come to pass? The advocates of free trade did not miscalculate the more immediate consequences which would result from carrying out the principle. The average price of bread was considerably reduced. Many articles of general consumption, such as tea and sugar, became much cheaper, and there was an extension of our commerce and trade which must have exceeded all expectations.

"In a few years the exports and imports of this country were nearly trebled. Almost simultaneously with the passing of free trade another agency was brought into operation calculated greatly to promote national prosperity. The development of the railway system then commenced, and the greatest improvement in the means of locomotion that had ever been introduced was rapidly spread over the country.

"If any one, a quarter of a century since, could have foreseen all that was about to take place; if he could have known that trade was soon to be trebled, that railways would be taken to almost every small town in the kingdom; would it not have appeared almost incredible that all these favourable agencies should have produced so little effect, that it may now be fairly disputed whether the poverty of the poor has been perceptibly diminished? There has, no doubt, been an unprecedented accumulation of wealth, but this wealth has unhappily been so distributed that the rich have become much richer, whilst the poor have remained as poor as they were before. * * *

"The leading characteristic of our industrial system is that the three requisites for the production of wealth are provided by three distinct classes.

"The land is supplied by the landowner, the capital is provided by the capitalist, and the labour is furnished by those who work for hire.

"Between these three classes there is the antagonism of opposing pecuniary interests; the landlord tries to obtain the highest rent for his land, and the capitalist endeavours to obtain the greatest amount of profits and to pay the lowest wages; the necessary capital and labour are supplied by two distinct classes, who, instead of being united by the bonds of a common pecuniary interest, generally act towards each other as if they were parties contending over a closely contested bargain.

"The employer strives to buy labour as cheaply as possible; the employed equally endeavour to sell it as dearly as possible. The struggle often ends in bitter hostility: an industrial war is waged which paralyses trade, and inflicts incalculable loss, not only upon the parties engaged in the dispute, but upon the whole community. The evil, however, is not simply confined to the mischievous consequences which ensue from an open rupture. There must constantly be a deadening influence depressing industry, as long as that antagonism of interest continues which now exists between employers and employed. Men cannot be generally expected to put forth their best efforts unless they feel that they have a direct advantage in doing so."

If we now glance at our National Income I think it will clearly appear that it is not in our case from the natural tendency that population has to increase faster than the means of subsistence that all are not well off, but from the mode in which that income is distributed.

The gross National Income Mr. Dudley Baxter estimates at £814,119,000. The upper and middle classes he estimates at 2,759,000, with independent incomes amounting to £489,474,000, and 3,859,000 dependent upon them—in all 6,618,000. In this class we have nearly three persons dependent to every two with independent incomes.

The Manual Labour Class Mr. Baxter estimates at 10,961,000, with an income of £324,645,000, and with 12,130,000 dependent upon them—in all 23,091,000. This class is almost equally composed of earners and non-earners.

The total income of the two classes just mentioned is equal to affording £27 per annum to every man, woman, and child in the kingdom, or if the population were divided into families of five persons, £135 a year to each family. As the income is at present divided, one-fourth of the people possess half as much again as the other three-fourths.

Again, of the first-named class 8,500 persons have an income of £126,157,000, and 46,800 have an income of £1,000 a year and upwards. The £100 line divides the total Income into two nearly equal portions. 1,262,000 persons, with £100 a year and upwards, have a total income of £408,000,000, while 12,458,000 have £406,000,000.

One-third of the total income is derived from Capital; two-thirds from Profit and Earnings.*

The real income of the Kingdom, however, can only be its annual produce in Agriculture and Manufactures. The Distributors of all grades and the Exchangers must be sustained out of this fund. In fact every person who does not produce either by head or hand must be kept out of it. The gross income of the Kingdom is thus very much reduced, for a large part of the population is engaged like two 'cute American boys who spent the day in swopping jackets with each other, and although they had nothing to begin with, left off with a dollar a-piece.

Mr. Baxter estimates the original earnings of productive income from Agriculture and Manufactures (the fund in which all the national income is at first received) at only £480,000,000. Partly original earnings, and partly second-hand paid out of original earnings, he estimates at £196,000,000; and the incomes of the non-productive classes, entirely second-hand, and paid out of original earnings, at £138,000,000. Here then we have more than a quarter of our annual income consumed by non-producers. The Army and able-bodied paupers ought to be producers,

* Mr. Taylor, F.S.S., read a paper at the British Association (1871), in which he tells us that "The total number of labourers in England and Wales of all classes living on weekly wages and working with their hands, is, including their families, 8,144,000, less than half the population. Of these, 1,178,000 are skilled artisans, or say 200,000 grown men; 4,009,000 are half-skilled artisans, or say 800,000 grown men; and 2,057,000 agricultural and unskilled labourers, or say 600,000 grown men. The average earnings of a skilled man range from £60 to £73 a year; of a half-skilled man from £46 to £52; and of an unskilled man or agricultural labourer from £20 to £41. The total earings of the working classes are £276,000,000 a year." From what sources this information is derived I do not know.

*i.e.*, self-supporting ; and this is now beginning to be recognised.

"The produce of labour," says J. Stuart Mill, "is redistributed at the present time in an almost inverse ratio to the labour supplied: the greatest return falls to the lot of those who never work: after these, to those whose work is only nominal, and thus in a descending scale, wages are reduced in proportion as the labour becomes more onerous and more disagreeable, until at last that which is the most fatiguing and pernicious to the body can scarcely secure with certainty the acquisition of the immediate necessities of existence."

The evils arising from this great inequality in the division of property are—

1st. The unnecessary increase of the non-producers in useless servants, &c.

2nd. The diversion of the industry of the country to the production of useless and even pernicious luxuries, and, as we have seen, the great fluctuation in the demand for the luxuries, and all the privation by which it is accompanied. A recent writer truly says: "The consumer of luxuries, who forces the capitalist to select unproductive employment, is responsible for blasting so much capital with the curse of ultimate sterility." (*Contemporary Review*, July, 1868.)

3rd. A false style and standard of living. As Political Economists aim at finding work and not rest, so in our social economy we try how *much* we can live upon instead of how little.

All our laws and institutions have tended to foster this inequality of condition—this unfair division of the joint produce of industry. If the Manual labourer instead of the Capitalist had made the laws, no doubt they would have been more in his favour.

Godwin, in his "Political Justice," puts the evils of this great inequality of wealth very strongly, and it has greatly increased since his time; he says: "And here with grief it must be confessed, that however great are the evils produced by monarchies and courts—by the imposition of priests and the iniquity of criminal laws; all these are imbecile and impotent compared with the evils that arise out of the present system of property. * * Excessive inequality of property brings home a servile and truckling system by no circuitous method to every home in the kingdom. Observe the pauper fawning with abject vileness on his rich benefactor. Observe the servants that follow in a rich man's train, watchful of his looks, anticipating his commands, not daring to reply to his insolence—all their time and their efforts under the direction of his caprice. Observe the tradesman, how he studies the passions of his customers, not to correct, but to pamper them—the vileness of his flattery, and the systematic constancy with which he exaggerates the merits of his commodities. Observe the practice of a popular election, where the great mass are purchased by obsequiousness, by intemperance, by bribery, or drawn by unmanly threats of poverty and persecution. * * The existing system of things directs all the efforts of mankind to the acquisition of wealth. * * The rich man stands forward as the only object of general esteem and deference. * * On the other hand, look at the peasant and labourer, working till their understandings are benumbed with toil, their sinews contracted and made callous, being for ever on the stretch—their bodies invaded by infirmities and surrendered to an untimely grave. * * The spirit of oppression, the spirit of servility, and the spirit of fraud—these are the immediate growth of an extreme inequality in the distribution of national wealth; and other vices become inseparable, envy, malice, and revenge, not to speak of violent offences against life and property," &c.

Partly to remedy this deficiency in more direct representation, the working man has been driven into electing Parliaments of his own in the shape of International Associations, Trades' Unions, &c. Laws have hitherto been made for the protection of property, and few, if any, for the protection of labour. The Wage Class have been obliged to make these for themselves, and the penalties they have inflicted for disobedience to them have sometimes been very severe, and not in accordance with the law as made by the Capitalist: witness the outrages that have constantly taken place at Sheffield and elsewhere, but which the workmen themselves are inclined to regard as only just punishments for the attempt to reintroduce "competition" among themselves, and thus take the bread out of others' mouths and lower the general rate of wages.*

* Sir W. G. Armstrong, in his Address, as President of the Economy and Trade Department, at the Social Science Congress (1870), says: "While we bow to labour as forming the essence of value, we must admit, as a consequence, that the price of labour determines the price of commodities. Now, whether commodities are raised to a fictitious price by protective enactments of the Legislature, or the protective restrictions of trades' unions on labour, the detriment to the public is precisely the same." At the present time one-third of the total Income or Produce of the Kingdom goes to Capital, and two-thirds, as we have seen from D. Baxter, to Profit and Earnings; so that if the public—as is too generally considered—consisted only of the two first classes, the Capitalist and Profit Owner, what Sir W. Armstrong says may be true, and it may be to the detriment of the public, but surely a little more may be given to the labourer without any detriment to the community at large. It is true that if the general rise in commodities were equal to the rise in wages, so that the increased wages would buy no more than before, it would be to the detriment of everybody; but such surely need not be the case. Sir W. Armstrong also said: "There is another species of restriction commonly practised by trades' unions which is much to be regretted. I refer to the regulations made for the purpose of establishing a uniform standard of wages, without reference to different degrees of skill or ability, and also for the purpose of deterring the able and the strong from exceeding a certain limit in the quantity of work they perform. There is a feeling of brotherhood displayed in these regulations to which I am ready to do justice, but at the same time I must declare that it is impossible to conceive anything more subversive of wholesome emulation, more unjust to able men, or more opposed to the public in-

But in looking forward to a better system that shall more completely include the interests of all, we must not forget

terest, than rules of this nature. Struggle for superiority is the mainspring of progress. It is an instinct deeply rooted in our nature. It shows itself in the keen interest which contest of every kind incites in our minds, and in the homage which we render to success, even in matters of little moment. To what a dead level of mediocrity would our country sink if struggle for superiority were stamped out amongst us, and how completely we should fall back in the race of nations! Trades' Unions would pursue a more rational policy if, instead of repressing the ability and energy of individual members, they were to foster and encourage these qualities, with a view of organising co-operative manufacturing societies." This is very true. There are many of the practices of Trades Unions that cannot be defended. No doubt the establishment of "co-operative manufacturing societies" is the most effectual way to remedy the evils of which the workmen most complain. Still, we have a long transition state to pass through first before the people are prepared for that; and in the meantime I see no reason why a *minimum* rate of wages should not be fixed by the workmen themselves, leaving the skilled and talented to get as much more as they can or deserved. Oh, but say the advocates of the present system, it is contrary to all the principles of Political Economy to fix the rate of wages. But nature fixes the rate of wages, and why should not man? Nature fixes the rate at as little as a man can live upon, and whatever Political Economists may say about the "impolicy of protection," he cannot take less than that; and surely therefore, as the rate must be fixed, there is no reason why the law, or the workman, should not fix it at as little as he can not merely live upon, but as little as he can live upon comfortably. Suppose it drove some trades away from us; why then let them go, if they would not keep the people employed in them *comfortably*. But the Empire is large, machinery is powerful, and there is no real fear of people wanting profitable employment when the interests of labour are as much considered as those of capital have hitherto been. This fear is only a bugbear set up by the capitalist to frighten working men into conceding what they call "free trade in labour," and which "free trade" is the instrument by which the great inequality of distribution has hitherto been brought about and maintained. Sir W. G. Armstrong condemned the outrages of which trades' unions have been too frequently guilty, expressed an opinion adverse to patent laws, and said in conclusion: "I have thus briefly glanced the good that has come out of the past. It has worked well at least for the few, and, as Carlyle says, "Whoso has sixpence is sovereign (to the length of sixpence) over all men; commands cooks to feed him, philosophers to teach him, kings to mount guard over him—to the length of sixpence." * It has released, also, a minority from bodily labour, and has developed that thought and moral feeling upon which the more rapid advance of the world in the future depends. Charles Darwin, however, says: "If man is to advance still higher, he must remain subject to a severe struggle. Other-

at the impolicy of protection, whether applied to commerce, to labour, or to invention. In every case the principle is the same, and amounts to a restriction upon human power and energy for the benefit of a section of the community. I deprecate all interference with liberty of action, except in restraint of offences; and I maintain that the greatest good to the greatest number will only be attained by leaving the social world to the governance of natural laws." The "natural law" to whose governance Sir W. G. Armstrong would leave us is the law of "unlimited competition" among workmen themselves for wages; but if the workman had made the laws instead of the capitalist, among the offences requiring restraint would probably have been paying below a certain rate for work done, and I am not among those who think this would have been an impossible or unjust restriction. I have lived among a manufacturing population where a *minimum* list price for work has been maintained, which did not prevent the best workmen getting more than such regulated price, but it certainly had great influence in preventing the lowering of wages, and if the list had had the sanction of law as well as of custom it would probably have done much more.

* On the present system a man works for his life, that is, for a "living," and consequently every vacancy is immediately filled. No place is so low that there are not always many applicants for it. This necessity for living, like water, fills every social cranny. An "assistant" and successor to Calcraft, the hangman, is wanted; and 134 applicants are eager to prove their fitness for the post. Out of these, seven, it is understood, have been selected, from whom the choice will finally be made. It is generally believed that a man named Toll will be the lucky person. Toll is famous among his companions as "a clever pig-killer"—pig-killing, it appears, having been his vocation before he became a dockyard labourer.

wise he would soon sink into indolence, and the more highly gifted man would not be more successful in the battle of life than the less gifted." But we can make no great advance, either Intellectually or Morally, under great bodily labour. If the force supplied to us from food is used up in vital and muscular action, it cannot be used in thinking and feeling, and the brain does not grow and increase in power. The population of Greece under its much-vaunted civilisation consisted (530 to 430 B.C.) of 90,000 freemen and 400,000 slaves, who did all the bodily work. The consequence was, as shown by Mr. Galton, a greater development of talent and of illustrious men than the world has known either then or since. The free-born population was at liberty to devote itself to culture, art, science, literature, philosophy, and politics; but if, at that time, they had entertained any scruples about the rights of humanity as to slavery, and they had had to do a fair share of the bodily work, the world must have done without their schools of philosophy. Machinery might take the place of the slaves in the present day.

Necessity, under the present system, has been the great mother of invention, and it has been the foundation of all order, as well as progress. Prof. Jacoby, in the speech to his constituents at Berlin before alluded to, says: "Robert Owen, the founder of the co-operative system in England, met one day in the house of a Frankfort banker, the renowned statesman, Frederick von Gentz. Owen expounded his socialistic system and displayed its excellence; if union could but replace disunion all men would have a sufficiency. 'That is very possible!' replied von Gentz, 'but we by no means wish that the masses should become at ease and independent of us; all government would then be impossible.' This, gentlemen," says the Professor, "is in two words the social question of the present time! For Owen the enigma of the solution is, 'union.' Gentz indicates the source of the evil which opposes this solution, 'the spirit of domination among the privileged classes.'"

But this "spirit of domination" has hitherto been a necessity, as the people are not at present at all qualified to govern themselves, and all government would have been impossible if people had not been kept in order by the necessity for constant and regular work. How far the people are yet from being able to govern themselves, was shown at the last election. All power, by the Tory party, was placed in their hands, and they returned a parliament of millionaires—of the men most interested in the support of the present economical system. The town population in France would disfranchise the country population. Much greater intelligence and moral power must precede all other great changes in our social state, before we can safely dispense with the "domination of the privileged classes." The Positivists reject co-operation. Dr. Bridges, a well-known Comtist, writing of the difference between the Positivists and French Socialists, while recognising "an important affinity," says: "They rely upon legal means for the execution of their purpose. We rely, in the main, upon moral agencies; only in a very secondary way upon legal. In their view, society should be a vast co-operative association, engaged in a vast variety of enterprises, regulating them by universal suffrage and by a vast machinery of Committees. Positivists have always rejected the co-operative theory. We think it would result, could the experiment be tried on a great scale, in feeble production, in wasteful distribution, and in aggravated selfishness. To us the concentration of wealth in a small number of holders is absolutely necessary to its efficiency. With some exceptions, as in the case of English land,* it is too

* Dr. Bridges excepts the English land. Mr. Motley, in an address at New York, said: "20,000,000 of men live in England, 30,000 men own England. The pyramid stands on its apex." * * This number was taken from the Census of 1861, and has been repeated by Mr. Mill, but the controversy to which this has given rise shows that 150,000 would be nearer the truth.

much dispersed, we think, even now. What we desire is to create an atmosphere of opinion in which the possessors of wealth shall be regarded, and shall regard themselves, as social functionaries, their functions being first to regulate production, so that the existing amount of wealth may somewhat more than reproduce itself; and, secondly, to see that a share of wealth, sufficient to permit healthy life, falls to the lot of the producer. And wealth concentrated will more readily be made amenable to this amosphere than wealth dispersed. In bringing about this state of things, law is obviously powerless. Christianity has proved ineffectual. Our hope lies in the system of thought and life which Comte has promulgated—in the religion of humanity." *

Probably we shall have to pass through some such stage as this contemplated by the Positivist, as the working classes have neither intellect nor moral stamina to act without the aid of the classes above them. They have always hitherto failed when they attempted to go alone. The elements of disunion—jealousy of each other and difference in opinion—have always been the strongest.

It is clear that in taking into consideration the institutions under which man can be made most happy, we must include his Political and Social Institutions as well as his Economical.

As to Government, no doubt that which is best administered is best, but that it may be well administered it must be adapted to the people over which it is to rule, and the degree of civilisation at which they have arrived. The British Government is probably the most free in the world, and it is a growth, an evolution, a development out of the centuries of conditions in which a people fitted for freedom have been placed. It has been the fruit of the merest empiricism, of the experience that has adapted each reform to the immediate wants of the people. It has been the creature

* T. H. Bridges. *Spectator*, April 29, 1871.

of circumstance, without reference to principle. It has been formed as Nature forms everything that she intends to be stable, and it has probably by this means developed, or is about to develope, the best system of Government possible. Government in England has passed through the Monarchical and the Aristocratic or Oligarchical form, and is now evidently approaching the Republican. From a contemporary account of a levée, at the latter part of the 14th century, at the court of Richard II., we learn that "the king on state occasions gave orders for a throne to be erected in his chamber, and thereon he used to sit, after dinner-time, until the hour of vespers, so that all might look upon him. He spoke to none, but looked at each individual in turn. And whenever any one, no matter what his station, encountered the king's glance, he bent upon his knees and did obeisance." This phase of royalty has passed away, and "obeisance" is now paid to the golden calf, and not to the throne, and the head of the state rules, but governs only in theory. The government of the Aristocracy ceased at the Reform Bill of 1832, and the people are now supposed to govern themselves; but that also at present is only in theory. In England it is wealth—the money-power that rules, guided by the public opinion as expressed by the Newspapers: and this is as well, as the public are far too ill-informed at present to govern themselves. The simplest and most just form of government is the rule of the majority, with full freedom of discussion to turn that majority into a minority. The people are supposed by a majority to elect the wisest among them to rule them, these again are governed by a majority among themselves; but this real Republican form of Government can only be possible or desirable where the people are all well disposed and well informed, which is far from being the case in any country in the world at present. George Combe truly says: "The conclusion, in regard to the republican form of government, which I draw, is, that no people is fit for it in whom the moral and intellectual organs are not largely developed, and in whom also they are not generally and extensively cultivated. The reason is clear. The propensities being all selfish, any talented leader who will address himself strongly to the interests and prejudices of an ignorant people, will carry their suffrages to any scheme which he may propose, and he will speedily render *himself* a dictator and *them* slaves."*

Intellectual and civil freedom necessarily pass by successive stages from the one to the few, and thence to the many; and it was for the announcement of his discovery of this law in history that Prof. Gervinus, in Germany, in 1853, (whose death has just occurred, March, 1871), was imprisoned, and his book publicly destroyed. His "Introduction to the History of the 19th Century" was charged with having advocated opinions dangerous to the State, subversive of public order, and involving the crime of high treason.

Democracy must advance. We cannot put the clock of time back; we must go on; and this is becoming generally recognised. De Tocqueville says: "Everywhere the various occurrences of natural existence have turned to the advantage of democracy; all men have aided it by their exertions. Those who have intentionally laboured in its cause, and those who have served it unwittingly; those who have fought for it, and those who have declared themselves its opponents—have all been driven along in the same track, have all laboured to one end, some ignorantly and some unwillingly; all have been blind instruments in the hands of God." * * "Is it credible that the democracy which has annihilated the feudal system, and vanquished kings, will respect the *bourgeois* and the capitalist? Will it stop now that it is grown so strong and its adversaries so weak?"† And J. K. Shuttleworth tells us that "the tendencies to democratic changes

* "Moral Philosophy," p. 375.

† "Democracy in America," Introduction to the First Part.

are so obvious, and are so strongly indicated by the origin, history, and theory of our constitution, that they are in some form utterly irresistible."* But let those who would go too fast and those who would lose heart learn something by the fact that "1,500 years have elapsed in our history, and yet the theory of our Saxon Constitution is only partially realised"; and also that Freedom, a blessing to the civilised and enlightened, is a curse to the ignorant.

Our Social Institutions also will certainly require a radical change if *all* are to be made as happy as they can be, instead of merely an upper ten thousand.

The Political Economist, for "right" or "wrong", now reads "tending to increase" or "tending to diminish wealth", and this without reference to Health, Intelligence, or the mode in which that wealth shall be distributed. What the Economists seek is an ever-increasing fund for the employment of labour. When each shall take his share of the world's work it will be found that what is wanted is an ever-increasing fund for the enjoyment of rest.

The Queen is the head of our Social as well as our Political State, and although she has ceased to govern in the one, she is all-powerful in the other. Her Majesty has an annual income of £385,000, that she may keep Gold Sticks in Waiting, and Lords of the Stole and Bed Chamber, and other Court Attendants;† and every one, following this example, tries to keep as many "Gold Sticks in Waiting" and

* "On Civilisation," p. 11, *Psychological Journal*, January, 1860.

† We find on the Civil List: Hereditary Grand Falconer (Duke of St. Albans), £1,200; Lord Chamberlain (Viscount Sydney), £2,000; Vice-Chamberlain (Viscount Castlerosse), £904; Keeper of Her Majesty's Privy Purse, £2,000; Mistress of the Robes (Duchess of Sutherland), £500; Groom of the Robes, £800; a Clerk of the Kitchen, who receives £700; and a Chief Cook, at a similar rate, &c., making the grand total of salaried retainers 921.—(J. Charles Cox. *Spectator*, July 29, 1871.)

other things in proportion as they can. But the time is coming when we shall try how *little* we can live upon, instead of how *much* we can spend.

The *Spectator* (April 15, 1871), however, says, probably with great truth as regards the present: "We doubt very greatly whether the tone of English society would be much changed by the disappearance of a Monarchy which has ceased to exercise social influence, but as far as any change occurred it would probably be for the worse. The millionaire would become the aristocrat, and the financier would lead society, and the ideal of life, instead of being merely luxurious, would be luxurions and vulgar too. Our social hierarchy, bdd as it may be, does restrain that worship of wealth which is the foible of Anglo-Saxons, and which threatens in America to demoralise every section of the community.

There is no denying that under the present phase of civilisation our Social and Economic system does very much for the favoured few, but it might, under better arrangements, do as much for all. We have seen that even under the present system, where so much is wasted, there is enough for all were it differently distributed; the capitalist and manufacturer, that is, about one-third of the population, now taking two-thirds of the annual produce: that is, for the use of land, machinery, capital, for superintendence and liberty to work, for distribution and protection, the working man gives eight hours' labour out of every twelve. This is rather an unfair division, but there is no immediate remedy; the privileged classes are not likely to give up their privileges and their "spirit of domination" without a considerable fight for them; still, as the labouring class get more intelligent, they are likely to demand a fairer share of the joint produce. The only effectual remedy for this unequal division is for the workman himself to become joint owner or hirer of land, capital, and machinery; but to do this a much higher rate of intelligence and moral feeling is required than he at present possesses. It has been a great mistake to divorce the workman entirely from the soil, as is the case of the artisans of our great cities. A deterioration of the physical stamina, and an undue predominance of the nervous system, with the too great use of alcoholic stimulants and tobacco, have been the consequence. Health, and security from poverty and want, can only be obtained by people occupying as much land as will furnish them with the first necessaries of life. A factory or staple industry should supply other necessaries and luxuries; so that if the latter should go out of fashion, the people would not be reduced to absolute want in the necessary change. Associative Industry, or Co-operation, should include a farm, a factory, and a store. The profits of the producer and distributor would thus be included; and there might be no more non-productive labourers than were actually required. This would correct the great waste in our present system, and procure for each "nearly the same enjoyments as if he were the single lord of all." The farm should be cultivated by skilled labour, by the aid of machinery; and by a union of small farms, the very best and most expensive machinery might be made to pay. The factory and store might employ the women and children of those families who were employed upon the land; and, by a proper organisation of such factories, they might be made greatly to assist each other. We have illustrations of farms, factories, and stores working successfully on the Co-operative principle; but they are working separately. The next step must be to bring them together.*

* "The working classes are believed to hold together £70,000,000 sterling in societies, savings' banks, and so forth. The number of co-operative societies is computed at 1,000, with a membership of 250,000. The yearly trade done is reckoned at £10,000,000 of money. Mr. Nuttall states that he has analysed the balance sheets of 150 co-operative societies, whose business was £500,000 quarterly. Of the whole annual result, nearly £500,000 was in groceries, of which, how-

"The Co-operative system," says the present Bishop of Manchester, "would encourage the re-appearance of small

ever, not more than £74,000 had been supplied from their own wholesale warehouse. * * Nevertheless, the Wholesale Society, in Manchester, do with the Manchester County Bank, on good terms, a yearly business of £1,000,000; and the societies with which they transact business do a trade amounting to £4,000,000 a year. To show the close affinity already subsisting between co-operation and trades' unionism, it may be stated that in the towns with which the Wholesale Society have dealings, the Amalgamated Engineers have 140 branches."—("The Future of Capital and Labour," p. 448, by George Potter. *Contemporary Review*, February, 1871.)

An elaborate return about industrial and co-operative societies in England and Wales was issued towards the close of the recent session (1871.) The whole number of industrial and co-operative societies registered up to the end of the year 1869 was 1,308; sixty-seven new ones were placed on the roll last year, making together 1,375. Of that number 406 appear to have been dissolved, and 153 have neglected to send up their annual statements. The following statement embraces 749 societies, with their figures up to the end of 1870:—The total number of members was 249,113, the admissions during the year were 42,314, and the withdrawals 21,964. Hence we may say that the former were to the latter as two to one. The amount of share capital credited during the year was £763,059; the amount debited, £582,835; the total capital on the 31st of December, 1870, was £2,034,261; loan capital at the end of the year, £197,128; cash paid for goods during the year, £7,458,000; received from customers in the same period, £8,202,000. The average value of stock-in-trade during the twelve months, £912,000. Hence the stock appears to have been turned over nine times between the 1st of January and the 31st of December. The expenses were £335,327, or rather more than 4 per cent. on the money paid across the counter; interest on loans, £92,333; and allowance for depreciation, £42,817: total of the three items, £470,477. This brings the expenses up to 5¾ per cent. The entire liabilities of the 749 societies are placed at £2,403,902; the entire assets at £2,649,429: the balance over liabilities in favour of the societies would therefore be upwards of £245,000. The value of buildings, fixtures, and land belonging to the various societies is stated to be £962,276. The disposable net profit realised from all sources during the year is put at £555,435. The dividends declared amounted to £483,687, of which £16,523 was allowed to non-

farms, without the re-appearance of a class that did neither themselves nor any one else much good—that of small farmers,"* and perhaps the evils from occupation on land alone are as great as those arising from isolated factory labour. In a late number of the "*Revue des Deux Mondes*" we have an account of the rather lamentable mental condition of the French peasantry:

"The peasant lives in a state of isolation, which the nature of his task imposes upon him, and which has become habitual, almost agreeable to him. He goes through his day's work silent and alone. If he has a companion by his side he seldom experiences the need of interesting him in his thoughts. Meal time comes; he still eats in silence. In the evening, on his return home, he sits down weary and harassed in the chimney corner. Is he dumb, or does he avoid, as useless fatigue, the trouble of translating his thoughts into words? What does he think of during these long hours? what subjects can occupy him? In such a condition there is but one: the comparison of the fate he endures with that of the happy rich. Thus passes middle-age. Old age, afflicted and ill-cared for, follows; carrying its bitter lamentations from door to door, and repeating to those who labour, 'This is what you will be one day.' * * Dispositions born of suffering quickly lead to a certain degree of moral depravity. * * We find ourselves then confronted with a population which, in this century of commercial and industrial progress, is generally miserable, and, remaining plunged in profound ignorance, deteriorates every day by contact with hurtful influences."

Take this in conjunction with the well-known condition of our own rural population. Dr. D. G. Croly, an American writer, says:

"The agricultural poor of England are to-day the most debased of

members; £3,776 was devoted to educational purposes. Lancashire is the birthplace of successful co-operative societies; we say "successful," because the abortive societies established by Robert Owen and some of his disciples will be remembered. In the return 155 societies are registered for that county—many of them very large ones.

* *Contemporary Review*, February, 1871, p. 451.

any class in Europe—are the worst fed, worst used, and worst paid. This cheap agricultural labour lies at the very basis of the aristocratic features of English landed property, and of their whole tenant system. Let the emigration fever once reach this lowest stratum of English society—and it is reaching it—and a heavy blow will have been dealt at the great tenant farming interest of that country, and at the wealth of the large aristocratic landed establishments. A very small advance in the wages of English agricultural labourers will make the raising of wheat and of all the cereals an unprofitable business in that country. It has already to a great extent done so, and hence the attention which has been paid in the last fifteen or twenty years to the growth and development of superior cattle."

And Nathaniel Hawthorne, in the "Blithedale Romance," that exquisite little sketch of an experiment in co-operation in land alone, says (p. 76): "The peril of our way of life was not lest we should fail in becoming practical Agriculturists, but that we should probably cease to be anything else. * * The clods of earth, which we so constantly laboured to turn over and over, were never etherealised into thought. Our thoughts, on the contrary, were fast becoming cloddish. Our labour symbolised nothing, and left us mentally sluggish in the dusk of the evening. Intellectual activity is incompatible with any large amount of bodily exercise."

A radical change, such as is here projected, can be but of very slow growth; at present little beyond a first step has been made in the establishment of Co-operative stores, saving only the profits of the retailer. Of course all the old-world prejudices and privileges will be against it, but what is worse, the wage-class themselves are by no means prepared for it; they will require a great educational change, and a moral and intellectual advance, that very few of them have at present made. Government might greatly aid in facilitating this change, as the great majority of mankind are very timid, have few convictions of their own, are perfectly willing to follow, and, in fact, almost require to lean on persons in authority; but it must be a Government for the people, and not, as hitherto, for a class—a Government in which the interests of labour shall be as carefully considered and protected as the interests of capital have hitherto been.

All permanent change is a growth. "Time is required for great changes, and all changes are the results of the continuous operation of natural forces which evolve phenomena in a regular succession. Nothing endures which is not in harmony with this law. The improvement in the condition of any class, the growth of any institution—such for example as a representative governing body—the relative powers and privileges of different orders of society, all result from a resolution of the great forces which create and sustain society. It is impossible to overleap intervals of time, so that conditions of society which a philosophic observer may foresee to be possible and to be ultimately beneficial, shall exist otherwise than as the result of that irresistible development to oppose which is to enter into a war with nature. Thus, in our own history, no change has been permanent which has not been a logical consequence of other antecedent conditions, all proceeding from the same social forces, and has not also been in harmony with surrounding conditions." *

Most of the actions of the individual are automatic—that is, performed involuntarily, the result of ages of transmitted growth; and so far as they are voluntary they greatly depend upon habit, custom, and public opinion: a man does what he has been accustomed to do, or what he sees others do, and rarely has an original impulse. It is very wonderful how under these conditions every one, without question, goes to his daily toil, however hard or degrading, and how small a body of police are necessary for our protection and to keep the community in order. Released from these ties, we have French revolutions, and every one trying to set his individual

* "A Sketch of the Laws of Social Progress, illustrated by the Growth of the Freedom and Political Capacity of the Manual Labour Class in England," by Sir J. P. Kay Shuttleworth.

will above his fellows. Society is governed in the same way as the individual. If it once loses its automatic action—if it is forced into new circumstances and conditions, and is obliged to think and decide for itself—all is contention, disorder, and confusion. It is this which makes any advance by force almost impossible. Sudden revolutions are useless, except for destruction, and we can only build on the living foundations. In a time of revolution physical force is necessarily in the ascendant, and moral force in abeyance. It is well, in the present absence of a Science of Man, that we are not dependent for action upon opinion, but are under wiser guidance. It is this transmitted experience of all that have gone before, bottled up in us, that guides our actions. It is an infinite consolation, in the jangle of creeds around, to know that God rules; that the right thing in the long run is the strong thing, and that all the talk goes for little or nothing. The great father of Idealism, Bishop Berkeley, whose moral instincts and actions were perfect, who had "all the virtues under heaven," had a room in his house fitted up for the preparation of tar-water, which he believed to be a universal balm for every physical ill, and with which he physicked his family and poor neighbours, and by which they were no doubt as much benefited as the world at large was, at that time, by his philosophy. At the present time, on all great questions, all our great men are about equally divided. Look at what was the state of opinion on the American and French Civil Wars; on Governor Eyre and Authority; on the Franco-German War; and at present on Theology; on Politics; on Political Economy; on Free Trade and Communism. Our leading thinkers are divided, and we have almost equal *authority* on each side; and such differences must continue until we have a recognised Science of Man. But fortunately it is not opinion that rules, for, as we have said, all permanent change is a growth. Consequently in the change contemplated from Competition to complete Co-operation we not only have to induce the external "conditions," but we have to grow the structure or brain on which the greatly needed increase in intelligence and virtue depends. We may have a perfect theory, but not persistent and consistent action without this. The old form of Society is by no means yet used up, and all experiments towards a more perfect form as yet have failed. "Time is the great innovator," says Lord Bacon; shall we then imitate time, which innovates so silently as to mock the sense?"

---

Prof. Beesley, in an article in the *Fortnightly Review* for November, 1870, gives us an interesting account of the International Working Men's Association. He tells us that the taxable income of the country had increased in eight years 20 per cent.; that the co-operative movement was "a victory of the political economy of labour over the political economy of capital." But the workmen are warned that these one-sided efforts are not enough. "The lords of land and the lords of capital will always use their political privileges for the defence and perpetuation of their economical monopolies. The workmen of all countries must enter into close bonds of brotherhood, and wars must cease. It is to promote these two objects that the International has been founded."

In August, 1868, 120 societies, of middle and southern Germany, held a congress in Nuremberg, and affiliated themselves to the International. The North German societies were prohibited by law from connecting themselves with any foreign society.

A new party, calling itself Democratic Socialist, met in congress at Eisenach, in August, 1869, where 263 delegates represented societies numbering 150,000 workmen. The programme of the International was adopted; and, as the law prohibited corporate affiliation to any foreign society, it was arranged that each member should belong individually to the International, and that the central committee, having its seat at Brunswick, should act at the same time as a central committee for the International. These were the men, we are told, who, in September (1870), faithful to the programme of the International, denounced Count Bismarck's claims upon France, and were, for so doing, sent to prison, chained like common felons.

In Austria the working-class opened relations with the International. On December 13, 1869, at the re-opening of the Reichstag,

40,000 workmen assembled in the streets of Vienna, to make their grievances heard by petition to the Ministry. On the 21st the bearers of the petition were arrested. The funds of the workmen's societies were arbitrarily seized and confiscated. At the end of July (1870) the delegates who had been arrested in December were placed on their trial for high treason, the public prosecutor stating that the principles of the International Working Men's Association were considered treasonable in Austria. Mr. Oberwinder, who had been one of the Vienna delegates at Bâle, and Mr. Scheu, were sentenced to six and five years' penal servitude respectively, and the rest to shorter terms of imprisonment. On 50,000 or 60,000 persons meeting to protest, they were dispersed by the troops, several being killed and wounded, and many arrested. A similar war against labour, we are told, has been begun in Hungary, the famous constitutionalist, M. Deak, being reckoned among its opponents. So that it seems if workmen's "petitions" or "protests" are to be heard, they must take their muskets with them when they go to present them.

In France members of the International were sentenced to long terms of imprisonment for "belonging to a secret society."

The fourth annual congress of the International was held at Bâle, in September, 1869, and was attended by 78 delegates from America, Belgium, England, France, Germany, Italy, Spain, and Switzerland. At its rising it resolved to meet in September, 1870, at Paris. But Paris had in the meantime resolved to go to Berlin, and as Berlin had come to Paris to render this call unnecessary, this meeting probably has not come off.

Touching the principles of the International, we are told by Prof. Beesley that "certainly the large mass of English members do not trouble themselves much about theories of property. They devote themselves simply to raising their wages, and diminishing their hours of work, by means of unions, and, in my opinion," says the Professor, "they show their wisdom by doing so." Whether it is the same with the mass of continental workmen I will not venture an opinion. According to the leading members of the International—who are, it may be observed, for the most part German—wage-paid labour is destined to pass away, as serf-labour and slave-labour have passed away; and it will give place to associated labour, which ought to be developed to national dimensions and fostered by national means, and that where the working classes possess political power, it is to produce these changes in a direct way. Others think that "where Communists went wrong was in wishing to introduce social changes by political means, instead of trusting to moral means.

M. Mollin, a Positivist, consequently, at the Bible meeting, proposed the following resolution: "The Proletarians here assembled solemnly repudiate the employment of government action, in whatever form, for the establishment of social systems; they declare that governmental action should be reduced to the protection of the liberty of all, and that no doctrine ought to prevail otherwise than by perfectly voluntary acceptance, resulting from free exposition." This M. Mollin, as one of the Paris Committee of the International, had been imprisoned in Ste Pelagie the year before. Not only those who would vote with M. Mollin, but all members of the International, are Republicans, and all determined opponents of war.

From the meetings held in London by this party in favour of intervention in the cause of France, it is to be inferred that through France and French Republicanism is thought to be the nearest way to the establishment of those views; but I rather think the way will be through a united and educated Germany. The French are too versatile,—wanting in Reflection, Firmness, and Conscientiousness, and nothing good and lasting can grow in such a soil. Professor Robert Knox, the anatomist, writing about 1850, says: "In my first lecture, delivered five years ago, I said that the Celtic race does not and never could be made to comprehend the meaning of the word liberty. The opinion I gave had no reference to recent events, but was deduced from past history: the histories of '92, of '15, of '32—add now the events of '48 and '49, and say have I erred in the estimate I formed of this race? On four eventful occasions the supreme power has returned into the hands of the Celtic men of France. Never was the destruction of a dynasty more complete. What use have they made of this power? * * As a Saxon I abhor all dynasties, monarchies, and bayonet-governments; but this latter seems to be the only one suitable for the Celtic man." No doubt there is much truth in this. "To revolutionise," he says, "is Celtic, to reform is Saxon." On the other hand, the Germans are a much more solid and enduring race, their distinguishing characteristics being Reflection and Conscientiousness. The nation is armed, and with increasing enlightenment they will discover the right and know how, if need be, to defend it—the military captains becoming captains of Industry. The one thing a Republican and Democrat does not understand is obedience, liberty with him meaning the right of every individual to do as he individually chooses; but real liberty is only compatible with obedience to law. The drilling the Prussians are now receiving is the best possible training for insuring that unity upon which the greatest strength depends, and it is to Germany, not to France, that we must look for the inauguration of the new social and economic world in its most complete and perfect form; England slowly working its own way onward in its usual empirical untheoretical manner.

Since this was written we have had a revolt of the workmen of Paris, an attempt to "introduce social changes by political means," to "solve social problems by force," ending, if the principles I have laid down be sound, and my estimate of the race be correct, exactly as might have been expected. The ostensible object was, "the recognition and consolidation of the Republic, and the absolute independence of the Commune extended to all places in France," as stated in a State Paper; but no doubt the real rising was that of the Labourer against the Capitalist, and which had been some time waiting its opportunity. The true history of this movement would be one of the most instructive pages that the world has yet seen. We have all the most advanced opinions, Republicans, Communalists, Communists or Socialists, Positivists, to which must be added perhaps 50,000 roughs, all bent upon putting the world straight and bringing about the millenium, in their own way. The objects, no doubt, of the great majority were the highest that can be conceived—the regeneration of the law, and quite prepared were they to become martyrs in the cause; but probably the most notable thing was the construction put upon the movement by the English Press, the *Times* leading the way, as in the great American Civil War, in the path of misconstruction. "The catastrophe," says the leading journal, (May 31st, 1871,) "is astounding, if measured against the cause. Men do die, and have died in all ages, for religion, but never for a 'religion' like this. No God, no man, no faith, no hope—nothing but better wages and more pleasure. The principle of the revolution now quenched in blood was not a bit grander than that of a tailors' strike." The *Spectator* (June 3, 1871) says: "The test of fraternity of the Communists of Paris has been the visible willingness to recent proprietary arrangements, and they have been apt to think no one deserving of being treated as a brother who did not acquiesce in this narrow, not to say mean, test of brotherhood. On the whole, we should say, that the terrible ferocity of the strife, as compared with the apparent smallness of the aims fought for," &c.

Mazzini published an article in the *Roma del Popolo*, in which he warns the Italian workmen against the theories of the International. He says: "The doctrines preached by the leaders and principal members of the International are: 1. The denial of God, who is the only, eternal, and unshakeable Foundation of your duties and rights. 2. The denial of the Fatherland and the nation, namely, of the points of support which you may all employ to save your interests and those of humanity. 3. The denial of all personal property, namely, of every inducement to produce more than what is absolutely necessary for daily life. Property, when it is the result of work, represents physical activity, as thought represents intelligence. These few words will suffice to teach you that the International can in no way be of any use to your cause."

An eminent physician present during the siege, as reported by M. Sarcey, "expressed the opinion that one of the chief causes of the terrible scenes which accompanied the final suppression of the Communist outbreak was a contagious mental alienation. The minds of the Parisians, he said, were gradually unhinged by the privations of the siege. The revolt of the 18th of March gave the last blow to brains which were already shaken; and at length the greater part of the population went raving mad." * * "Look at the citizens," he said, "who did not take part in the insurrection; they are immovable and stupid, like men struck with paralysis. Yet they have on many occasions given proofs of uncommon vigour and courage; but the air of Paris is at this moment poisoned by deleterious miasmata which make some people furious madmen and others helpless idiots. I have seen the strongest intellects stagger under this pernicious influence, and I have myself frequently felt as if my reason were going."

No doubt there is considerable truth in this view of the matter, and it would have, as in the great Revolution of 1789, a very serious effect upon both the bodily and mental constitution of the next generation.

Besides Mr. Sarcey's "Siege of Paris," we have the testimony of an American, Mr. Nathan Sheppard, in "Shut up in Paris," (London: Richard Bentley and Son, 1871.) "The vain boasting of France," he says, "is the only serious habit of the people." His view of French democracy is given in the following significant extract:—"Republicanism cannot survive where there is not a certain degree of public intelligence, a certain amount of virtue, and a certain measure of self-reliance. The masses here have none of these. They have no confidence in one another, and have less to fear from a ruler of their own choice than from themselves in the experiment of ruling themselves. The worst ruler France can have is—'France.' In the United States and in Great Britain every man stands upon his two solid legs. Here every man leans against every other man, and all have an ineradicable desire to lean against some one man."

"If Germany," says Mr. Sheppard, "could have been conquered by the tactics which reduced the walls of Jericho, the entire army of the

Empire would now be encamped in the suburbs of Berlin." The effeminacy and thoughtless excitement of the citizens, according to Mr. Sheppard, were everywhere apparent. A revolution delights them, but it fails to make them serious. "Such antics and frantics, such grotesque contortions of rage, such gesticulating and perspiring, and shaking of fingers and brandishing of fists and hats, and such laughing and jesting too." They have not the least symptom, he says, of possessing a back-bone, moral or physical; they have no "grit".

Mr. Frederic Harrison, a well-known Positivist, gives us his opinion in the July, 1871, number of the *Fortnightly Review*, of the Commune, the concluding passages of which are well worth attention :—

"This great crisis had stated though it had not solved the social problem. What, in a few words, is this problem? It is this. In this complex industrial system wealth has discovered the machinery by which the principal, in some cases the whole, results of common labour become its special perquisites. Ten thousand miners delve and toil, giving their labour, risking their lives; ten masters give their direction or their capital, oftenest only the latter. And in a generation the ten capitalists are rioting in vast fortunes, and the ten thousand workmen are rotting in their graves, or in a workhouse. And yet the ten thousand were at least as necessary to the work as the ten. Yet more. The ten capitalists are practically the law-makers, the magistrates, the government. The educators of youth, the priests of all creeds, are their creatures. Practically they make and interpret the law—the law of the land, the law of opinion, and the law of God; they are masters of the whole social forces.

"A convenient faith has been invented for them by moralists and economists, the only faith which in these days they at all believe in—the faith that the good of mankind is somehow promoted by a persevering course of selfishness. Competition is, in fact, the whole duty of man. And thus it comes that in ten thousand ways the whole social force is directed for the benefit of those who have. Habitually, unconsciously, often with what they think is a religious sense of duty, they work the machinery of society for their own objects. In this favoured land, whilst the owner of the soil knows no other toil or care but that of providing fresh modes of enjoyment, the peasant, out of whose sweat his luxury is wrung, lives like a beast of burden, and dies like a dog in a ditch; whilst the merchant prince is courting society for a peerage, a thousand lives of seamen are lost, decoyed to sea in rotten ships; whilst mine owners can still paralyse the Legislature, a thousand lives are lost each year in pits, 'chiefly, it is said, from preventable causes;' and whilst fortunes are reared by ironmasters, a hundred thousand workmen are ground to the dust by truck. Let us reflect what is implied in this mere finding of the late Commission. One hundred thousand families in England are cheated, insulted, and oppressed by being forced to barter portions of their wages for some fraudulent equivalent in goods. Now all this makes up in gross that which they call in France 'l'exploitation des ouvriers.' They say that where in a common work labour is no less necessary than capital, and labourers are as worthy of the profits as managers, the system by which the gross result is appropriated by capital, and under which the self-indulgence of wealth soars to yet unimagined heights, whilst the area of misery, ignorance, and exhaustion sinks ever deeper, is a system which is doomed to end. And this their claim is good.

"Let us turn to the remedy they propose. The whole social force which so long, they say, has been directed by capital in its own interest, shall be directed by workmen in the interests of workmen. The laws shall no longer be made and administered so as to handicap the labourer in the race of industry. The power of the State shall step in to neutralise competition, and to restrain the selfish abuse of capital. The land, at any rate, they say, must be resumed by the State for the benefit of the whole community, and farmed on social, and not on proprietary, bases. Ultimately, in short, the whole existence of capital, and the ordering the lives of the community, must be subject to the will of social authority.

"Such is the faith which, in spite of its extravagance, has seized the foremost minds of the workmen of Europe, which in some form or other receives the devotion of a religious creed. Can any one doubt its strength compared with the conscious corruption of the opposite creed? Does the selfish cunning of competition in its heart think it can stand a social energy like this, with all its errors and all its dangers? Does a society which lives in its equipages, and toils only in amusement, match itself seriously with men who are ready thus to die for a cause however mistaken?

"The claim of capital to amass wealth by what means it chooses, and to spend it how and when it pleases, is so vile, the claim of the workman to have his part in the social result is so unanswerable, that in the end the issue is not doubtful. And since this social problem must some day be faced by all, it seems time for serious men to reflect what other solution remains. Communism stares them in the face; it grows and deepens. Whatever it may suit a journalist to say, no sensible man believes that the 200,000 men who voted for the Commune are bandits and fiends in human shape. They who think that such a story as that of the Commune of Paris is explicable on the 'miscreant' theory are unfit to discuss political questions. It has a great purpose, and it has great leaders. For every man who died on the barricades ten will spring up hereafter. The cry of Millière as he died, 'Vive l'humanité!' will not be unanswered. The bones of Delescluse may be burned in quicklime, but his spirit lives. He and his followers have a purpose. They have sworn that the 'exploitation of the workman' shall end; and end it must.

"They are not so clear about a substitute, but, so far as they have a scheme, it is Communism. There is but one alternative—the answer of Positivism. In one sense Positivism is akin to Communism, for it heartily accepts its belief in social reorganisation; in another it is the opposite of Communism, for it exalts instead of abolishes the exercise of property by individuals. Positivism teaches that the highest uses of society are best served by massing capital in aggregates, and by entrusting these masses to the free control of individuals. It teaches that the dignity of domestic life and of public action, that moral as well as intellectual energy, demand the existence of capitalists as a class. It shows that the highest functions of social life and the noblest powers of the character would cease to exist without the free control of capital. It insists on this freedom in no grudging or unreal spirit. It looks on capital simply as power, and the wise exercise of power as the noblest function of man, and the trustful respect of power as the most generous of human instincts.

"And if Positivism insists that Communism would dwarf and unman every side of human nature, by destroying the infinite sources of nobleness inherent in individual property, it asserts as strongly as Communism itself that individual property can no longer exist on its prevalent conditions. It insists that the use of property must again be made a religious duty—that capital arises from the combination of many efforts, and must in justice minister to the common wants. It would, by an education prolonged through life, teach the workmen of every grade to trust and aid the owner of capital, and the owner of capital to look on himself as the minister of a free community. It would train the rich to rely on their own resources, and compel them to use their full responsibility in so ordering the common industry that the fewest disasters and the least suffering befell the labouring community. Such is a picture of capital not extinguished, but moralised; not cut in pieces, but raised to new functions; not harassed by the fetters of law, but strong in the noble consciousness of public office. Let capital, shrinking from the fires and shambles of Paris, choose which of these two it will have. Communism is a not impossible future. Positivism is a not impossible future. The *status quo* is impossible. The alternative is Communism or Positivism."

Positivism is thus a benevolent and enlightened despotism, and it is probably more possible in the transition state that the few men could be found qualified so to govern, than that the many could be yet found capable of governing themselves. Otherwise, without the right men Positivism would appear only to present all the worst features of the present society in an exaggerated form, viz., a greater despotism based on larger fortunes; not but that the despotism of the many when in power has always hitherto exceeded that of the individual.

If we want an admirable illustration of what the Capitalist may do for the Workmen, when enlightened and so disposed, we must consult "Solutions Sociales." Par Godin. (Paris: A. le Chevalier, Rue Richelieu). M. Godin built a palace for his workpeople, giving them all the refinements and conveniences of modern civilisation, and the rents, with the profit on the Stores, paid him 6 per cent. for his outlay.

# CHAPTER VII.

## SUMMARY AND CONCLUSION.

Geological periods mark only epochs, not times.

Life has probably existed on this earth 100 millions of years; Man about 100 thousand.

The faculties, bodily and mental, which man possesses, have been gradually perfecting in every form which life has taken since its first appearance in the world.

So far as the history of Creation is written in the geological records, it would appear that God's mode of working is by evolution and development, by which all things are made to make themselves, and not by special creation. The whole world is one complete and living organism. "The tendency on the part of matter to organise itself, to grow into shape, to assume definite forms in obedience to the definite action of force, is, as I have said, all-pervading. It is in the ground on which you tread, in the water you drink, in the air you breathe. Incipient life, as it were, manifests itself throughout the whole of what we call inorganic nature."* Every atom has its likes and dislikes, and moves with instant precision to its place appointed, and the presence of mind alone, as Sir John Herschel says, explains the whole difficulty. Mind has its laws as well as what we call Matter, and when we exercise the highest powers of our intellect we act in accordance with certain laws of thought. In the action of the Supreme mind we may expect the same to hold, and God's action therefore is not an interference with law. He acts as necessarily in accordance with the laws of his own

* "Fragments of Science," p. 115, by Prof. Tyndall.

being as we do in accordance with ours.* The artificial distinctions and divisions that have been set up between inorganic and organic are becoming daily less and less visible in the light of science. There is no necessity to bring the torch of life from other worlds, in meteoric stones or otherwise: the whole world is alive. The conditions were evolved and individual life was born, although it may not be in man's power to beget such conditions, and Life now proceeds only from Life.

* The Rev. Prof. Caird, in a Sermon preached before the members of the British Association, at Edinburgh, August 6th, 1871, says: "For the more men advanced in intelligence, the more clearly did they begin to see that it was only a vulgar necessity of thought which identified personality with changefulness, and arbitrariness with paroxysmal acts and special interferences." "What, then," he asked, "to a thoughtful observer would be the kind of phenomena, the aspect of things and events, which would look most like the signs of personal thought and will in nature? Surely those phenomena and that aspect from which the indications of anomaly were most completely banished, and throughout which, from beginning to end, reigned calm and changeless order, unbroken sequence and continuity, the majestic presence of power and law."

Mr. C. Staniland Wake, in a paper read before the Department for Anthropology, August 8th, said: "It was simply impossible to understand or to reason with any person who entertained a notion of a creation of something out of nothing. If they could not have something out of nothing, they must have evolution of some kind. It had been said that he had taken Pantheistic ground. Well, he believed that nature was God, a personal God, and that all nature had been evolved."

"Dr. Cairns: Does the gentleman hold that nature is a personal God?

"Mr. Wake: Yes.

"Dr. Cairns: Thank you."

There seems a considerable gulf to fill up between the Rev. Drs. Caird and Cairns. The latter appears to agree with Mr. James Hinton ("Man and His Dwelling Place"), "that all our conceptions are based on the implied postulate that the world is as it appears," but he does not equally agree with him that "the advance of knowledge consists in the substitution of accurate conceptions for natural ones."

Sensibility, or Mind, with animals has been always connected with the Nervous System, and has developed and improved as the Nervous System has been developed and improved in its transmission from body to body in the chain of being.

Man differs from other animals, not in kind, but in degree only.

The "plan," or object, in Creation would appear to be to produce the largest amount of enjoyment without respect to species, or individuals, or man exclusively. "Freedom" is only one of the means towards this end, enabling all the faculties upon which happiness depends to be brought into activity.

Man knows only his own consciousness, or rather his consciousness comprises all his knowledge. External force, acting upon a portion of the structure of the brain and nervous system, creates within him the "World" as he conceives it in his intellectual consciousness, with the ego, or sense of personal identity; passing through other portions of the brain, it creates those likes and antipathies upon which the moral world is based.

Each creature has thus created within it a world of its own, according to its susceptibility to impressions from without, which susceptibility depends upon the specific nature of its nervous organisation.

These "modes of action" of force without give us no real knowledge of things without. We know only how we are affected by them—that is, we know only the consciousness they create within us. All is illusion: colour, on which the beauty of the world so much depends, is a feeling or idea, a nervous sensibility; and so are the forms which we believe to be around us mere ideas. Nervous force only is transmitted to the brain, and the brain forms the image. Without us is we know not what, except that it has the power to produce these feelings and ideas, and that all power is the power of God.*

* The opposite or Realistic view is thus well expressed by Prof. Donovan: "Man is brought into relation with external objects by means of Faculties, each one of which corresponds with a special property of the object. Objects have 'Form;' man has an Organ of Form, by means of which he perceives this attribute. Objects have 'Size;' man has an organ by means of which he takes cognisance of Size or Magnitude. Objects have 'Weight,' or gravity; man has a corresponding organ; and so on of 'Colour,' 'Order,' and 'Number.' Each has its proper Organ, and these organs are placed in the superciliary ridge or eyebrow." Of course *our* belief is that the organs create the ideas to which we have given these names of Form, Size, Weight, Colour, Order, and Number, without at the same time giving us any real knowledge of the properties of objects themselves, for it is impossible to conceive that an external force—the property or quality of the object—can have any likeness or resemblance whatever to an internal feeling or idea. All we *know* is the idea; from that we *infer* the quality, and the substance. But the quality is force, and we logically infer that it is the force of something, but of what we do not know. We may call it substance, matter, mind, spirit, but these are merely names for the unknown.

How completely our men of science have set things up on the wrong end,—for the external is nothing but "possibilities of sensation,"—is well illustrated in the Presidential Address in the section A. at the last meeting of the British Association (1871.) The President says: "I conclude by quoting some noble words used by Stokes in his Address at Exeter,—words which should be stereotyped for every meeting of this Association: 'When from the phenomenon of life we pass to those of mind, we enter a region still more profoundly mysterious. * * Science can be expected to do but little to aid us here, since the instrument of research is itself the object of investigation. It can but enlighten us as to the depth of our ignorance, and lead us to look to a higher aid for that which most nearly concerns our well being.'" No wonder Anthropology, or the Science of Man, is squeezed into a mere department of Biology at the British Association when this is the kind of *stuff* the Association considers desirable to have stereotyped every year out of deference to a supposed religious public opinion. The only knowledge we really have, all that can be of any real importance to us, is of our ideas and feelings, and these are immediately connected with the brain; and yet the members of

The ideas thus created are not intended to give us real knowledge; only to guide us towards the objects of our being, upon the attainment of which objects our happiness depends. Matter and spirit are only "the names for the unknown and hypothetical cause of states of our own consciousness," and all controversy based on their supposed essential difference is so much mere babbling about that of which we necessarily know nothing.

the British Association generally do not know whether they have any brains or not, (neither does anybody else), and the Science of Mind, based upon Cerebral Physiology, is to this learned body "profoundly mysterious," and any study of it can but "enlighten it as to the depth of its ignorance." A very likely consummation, and most devoutly to be wished, at least as far as this most important department of all Science is concerned! A knowledge of the phenomena of mind, we are told, can only be attained through supernatural aid: in this knowledge the British Association is no further advanced than the world was more than 2,000 years ago, and there it seems likely to remain. It shows how completely current superstitions stand in the way of all progress.

From Prof. Tait's and Prof. Allen Thomson's allusions to the phenomena of so-called Spiritualism it is probable that neither of them have heard of Mr. Crookes's careful experiments on the new (Psychic) Force; or do they consider this gentleman, Dr. Huggins, and Sergeant Cox mere impostors? The bigotry of orthodox science I have always found equal to the bigotry of orthodox religion; but in this case the men of science might even take a lesson from the Clergy. The Bishop of Gloucester and Bristol would have made a better President for their sections than either Prof. Tait or Thomson, for he does seem to have some little intimation of the new truths that are coming upon us. The Bishop says, in an admirable paper on Death in the *Contemporary Review*, for August, 1871, p. 61: "Such terms, vague as they really are, as *medicatrix naturæ* and the like, are all indications at least of the widely spread persuasion that forces are at work within us which are essentially and qualitatively different from any of those with which experimental philosophy has yet come in contact." The British Association is treating these new "forces" just as orthodox science always treats new truths.

"The man of science himself is fonder of glory, and vain,
An eye well practised in nature, a spirit bounded and poor."

—Tennyson.

On the principle of evolution the future necessarily grows out of the past and present: that is, "everything that exists depends upon the past, prepares the future, and is related to the whole;" according to the recently-discovered law of the Conservation of Energy, "each manifestation of force can be interpreted only as the effect of some antecedent force, no matter whether it be an inorganic action, an animal movement, a thought or feeling."

Consequently, all actions being equally necessary—all equally the effect of some antecedent force, there can be no intrinsic difference between them, the only difference being one of arrangement. Good and Evil are purely subjective—that is, dependent upon the way in which our sensibility is affected by things without. Where we have pleasure it is called Good; where we have pain, Evil. Pleasurable sensation attends the legitimate action of all our faculties, whereas pain or suffering is not the legitimate object of any part of our organisation. Praise and blame, reward and punishment, are not a recognition of any intrinsic difference in actions themselves, but of our wish to produce one class of actions rather than another, as more agreeable to ourselves. They are intended only as motives to action.

Responsibility consists in our having to bear the natural and necessary consequences of our actions. The supposition that our responsibility consists in our liability to so much suffering for so much sin or error, if not in this world, then in another—that *justice* requires that if we sin we must suffer—however ancient, is an altogether groundless notion. The object of pain or suffering is reformation, and any pain or punishment that has not that object, any suffering in excess of that, would be objectless and mere revenge. Every sin contains its own atonement in the pain or penalty attached to the natural consequences that follow it. We do not require an *exact* apportionment of so much suffering for so much sin, which, it is said, only an infinitely wise Judge can deal out in another world; all that is required is so much pain or punishment only as will prevent a repetition of the offence. It is not the sense of justice that begets the persuasion—the strong feeling within us, that evil will be punished, but revenge; d—— you! is not "moral purpose;" it is the language of destructiveness, not conscientiousness, and the old Hebrew conception of so much suffering for so much sin is altogether a barbarous notion originating only in a savage state. That Retribution would not be *just* which included more punishment than was sufficient to correct the offence and was therefore good for the offender.

If, as Quetelet says, "Society prepares crime, and the guilty are only the instruments by which it is executed," the strict demands of justice would require that the sinner, not the saint, should be made happy in another world, because the sinner, having been *made* to "dishonour" in this world, has been the most unhappy here, and requires compensation.

Morality is the law, or the rules and regulations by which men may live together in the most happy manner possible. It expresses the relation of man to man. The feelings that induce us to seek our own well-being are necessarily strong, or we should cease to exist: so strong that there is always a tendency to gratify them at the expense of others; whereas the moral feelings that unite us to our fellows, that bind us up in humanity, are not so strong; we praise them therefore to induce their action; but there is no intrinsic difference between them, as is usually supposed. All our feelings are equally necessary for our well-being, and are equally pure and good, and all equally bad when in excess. We must "do justice, love mercy, and walk humbly with God," but we must not neglect our social duties and our daily work to attend to the interests of humanity. What is wanted is the proportionate activity of all our faculties, and for such activity there is full scope in this world, and we do not require to wait for another to bring every faculty we possess into full activity, as some religious people affirm. A man may be happy as a pig, or as a peacock, happier still as lover, father, or friend; but he is happiest of all when is added to these the full exercise of the moral sense—of the duties that make him one with his fellows. Every feeling has its reflex action—its reaction. If he is kind, others are kind to him; if he is just, others are just to him; if be gives full measure, it is measured to him again. If also he seeks God, he will find him in the harmony and beauty and sunshine and enjoyment around him. All these feelings do not differ in kind, but in degree only, although what is called "moral purpose" is only to be found in man; they all make him more or less happy, as they are intended to do.

As Morality is the relation of man to man, so Religion is the relation of man to the ever present Spirit of the Universe. God is not more present in one place than another. "The true doctrine of omnipresence is that God re-appears in all his parts, in every moss and cobweb; thus the Universe is alive."* "There is but one infinite substance, and that is God. He is the Universal Being, of which all things are the manifestation."† Science is the record, the classification and arrangement of His "mode of action." All His laws are equally sacred, and upon man's knowledge of them and obedience to them his happiness depends. All, therefore, ought to have equally the religious sanction. We want something to worship, and from the lichen to man, through all the varied chain, we find all worthy of our veneration and adoration; but to make a Religion of Humanity is going back to only another form of fetishism. Humanity is only a small element in either our admiration or enjoyment. We may be taught also to admire all that is great and good in Humanity without making a Religion of it. "The just may be had in everlasting remembrance" without our worshiping them. When once we worship,

* Emerson. † Spinoza.

justice and due discrimination is at an end. The feeling takes captive the reason. We have such a thing as Bibliolatry where people worship the Bible, and "that blessed word Mesapotamia" becomes the greatest source of comfort to the old women. The devout worshippers of Shakspeare defend his puerilities equally with his surpassing excellencies. We always begin with depreciation, and end with undue appreciation. It is an ever-present God, as seen by the lights of our latest knowledge, that we are alone safe in worshiping. All worship of the past ends in retrogression and a stereotyped civilisation.

Death is a mere correlation of force; each cause and effect being a new Life and a new Death: one form gives up its energy, latent or active, that another may live: it is the form, appearance, or phenomenon only that passes away. There is no real Death; no such thing as annihilation. Force, which we call Spirit, is persistent or indestructible. What we call Death is the mere grinding up of old and useless bodies to make into new and more perfect ones. "All death in nature is birth—the assumption of a new garment to replace the old vesture which humanity has laid aside in its progress to higher being." If we have gone wrong—and we all have—and use has become second nature, that is, habit has become organic or constitutional, the old body is broken up and starts again in some new form without such impediment to right action. By such a process, and for such purpose, death sunders the immaterial, as it is called, from the material.

Our Social and Economical System causes great inequality in the distribution of wealth. With better arrangements, by the aid of Science and Machinery, all might be well off. The great question of the day is this relation between Capital and Labour, and it can only be ultimately settled by the Capitalist and Labourer becoming one; by Land, Capital, and Labour, working together on the Co-operative principle, in which all employed shall have a proprietary interest in the concern, and take a fair share according to that which all are jointly instrumental in producing. The sun does the greater part of the work,* and it has been made to work through

* Philosophers now recognise but one force and its correlatives, or various modes of action. For practical purposes, however, it is found desirable to recognise two forces—sun power and earth power. By the silent and imperceptible force of the sun, calculated by Leslie to be two hundred thousand times greater than that of the combined exertions of the whole human race, water is raised in vapour, which gives out the same amount of force or earth power in finding its level on its way back to the ocean. The air is also rarefied by the sun, which in finding its equilibrium produces a force not much less; and we have thus wind and water power—the sun and earth power balancing each other—which, if it could be economically applied, would be sufficient to do all the work of the world. But Wind and Water have given place to Steam, the power of which, as a force that can be generated at will, is again but bottled sun-shine. Millions of years ago, perhaps, the sun divorced the oxygen from the coal or carbon, which upon their re-union in what we call combustion creates all the power of steam. The same sun power now at work in plants separates the carbon from the oxygen, which upon their coming together again in the animal system is the source of all animal power.

"It is certain," says Prof. Tyndall, "that the sun has been chilled to an extent capable of being accurately expressed in numbers, in order to furnish the power which lifted this year a certain number of tourists from the vale of Chamouni to the summit of Mont Blanc."—("Fragments of Science," p. 437.) People were beginning to suspect that too much had been taken out of him in this way, this cold summer (1871), and they were thinking of whether it might not be desirable to forbid such expeditions, until they were reassured by a return of heat in the autumn (August) of 87 deg. Far. in the shade in London. Now this force of the sun has been duly recognised, it is proposed to use it directly, as well as indirectly. Mr. Ericsson, a Swede, resident in New York, so well known for his efforts to improve the steam-engine, has now turned his attention to the grand question of utilising the sun's rays; and he states that he has devised apparatus by which the heat may be concentrated and used for steam or air-engines. In other words, the solar radiation accumulated on a space ten feet square and perpendicular to the sun's rays will

machinery almost exclusively in favour of the owner of the machine; and this it is that has caused the great inequality of distribution.

develop somewhat more than one-horse power. Hence the sun-shine that falls on the roofs of Philadelphia, he calculates, would keep going 5,000 steam engines of twenty-horse power each. This suggests many reflections; it is at once clear that all the gloomy forebodings about the exhaustion of our coal may be entirely dismissed. The Laputians it is, I think, that Dean Swift represents as being employed in extracting the sun's rays from cucumbers; it appears we need not resort to this, but every man may have a little apparatus—of one-horse power—on the top of his house, which, with machinery judiciously applied to every room, may do all the drudgery and household work of the establishment. This one-horse will not require to be fed on anything but sun's rays, and daylight always contains a sufficient supply of them. It must not be supposed that wives would then go out of fashion; they might, on the contrary, for the first time take their proper place. If man is the "power" in the world, woman is the "beauty"—the æsthetic is properly her department.

But such a power need not be confined to domestic uses. In India, in the temples of Buddha, the prayers are done my machinery, which is kept going night and day. Here it has been calculated that a one-horse sun power, or even less—a one-ass power—on the top of our churches, would probably be enough to preach all the sermons. The ideas in our ordinary sermons are so few and simple that they might readily be set to machinery, and ground out every Sunday for the benefit of those very young people who are not yet familiar with them—thus releasing a considerable amount of ecclesiastical force that now undoubtedly might be applied to better purposes. Good sermons may be purchased in lithograph at from 5s. a dozen. I am rather surprised, therefore, at the following, which I take from the *Pall Mall Gazette*, April 28, 1871:—

"Clergymen with a small stock of sermons frequently speculate too boldly on the shortness of their hearers' memories, but they probably seldom carry the practice so far as the reverend gentleman whose sermons have been subjected by one of his congregation, 'a great statist and an old resident,' to the following minute classification. A friend of the 'great statist' thus reports the result of his inquiries:— 'His theory is that during thirteen years of the present incumbency, the general stock of sermons possessed by the vicar has been preached

We have now only to draw the legitimate deductions from the above, and to see how we may best further nature's plan in the production of the largest amount of happiness in the world. As far as we are individually concerned we shall perhaps do this by losing sight of happiness altogether, and not taking it into account at all. The objects of our desires are all closely marked in our constitution and organisation, and not one of these desires is the desire for happiness. It is upon this partial truth that the anti-utilitarian rides off

forty-eight times, or, as he says, has made forty-eight revolutions, and that they are already far advanced in the forty-ninth cycle. He thinks that when they have gone through fifty revolutions they wi l be tolerably well worn out—seeing that they were ancient when they came into the possession of the present owner. He considers that he has accurately fixed the date of their composition, which he believes to be between the thirty-fifth and fortieth years of George III., or about seventy-five years ago. Two of them, which are known as the "Astronomical Sermons," he says, are taken from Derham's "Astro Theology," published in 1786, and abound with the errors of that date. Another is supposed to have been written on the occasion of the earthquake at Lisbon, and is known as the "Trembling Sermon," being suited to occasions of public calamity. It has already done duty on the occasions of two revolutions in Spain, twice for commercial panics in England, once for the Crimean war, and recently for the Franco-German War, with a few other similar occasions. They are taken from some early numbers of the *Church Magazine*.'"

We have not yet been able to tame the Lightning; it has proved itself at present a little too eratic in its course for daily use. A correspondent of *Nature*, however, draws attention to an important discovery made by an American, which bids fair to supply the want of a portable motive power for machinery. At the last exhibition of the American Institute, it seems, there was shown an elliptic lock-stitch sewing machine driven by an electric engine small enough to fit into a common hatbox: "A series of eight magnets are set on the periphery of a circle, and around these revolves an armature of steel which is continuously propelled by the magnetic action, and thus operates the machinery that moves the needle. Connection with this motor is had by means of a small slide within reach of the operator, at whose will the current may be cut off entirely or the speed of the needle graduated as may be desired."

triumphantly. We love our wife, children, and friends—not happiness, but the love of them produces happiness.

We have seen that from the commencement of animal life in the world, pleasurable sensibility has increased with the increase in size of the nervous system, and intelligence and wants have increased with the happiness dependent upon their gratification with increased speciality of structure. This is still the law. What we want to grow is more brains; not at the expense of any other bodily powers, but proportionate to them: a sound mind in a sound body. But we not only want more brains, but we want those parts of the brain especially developed that are connected with the highest attributes of man—the intellect and moral feeling. In fact, we want to improve the breed of men, our attention hitherto having been almost exclusively devoted to dogs and horses, sheep and short-horns. We can produce almost any variety of plants, flowers, and fruits, and almost any form in animals; but all such laws are entirely ignored as far as man is concerned. Providence is supposed to take this department of science entirely into its own hands. We have only to procure a marriage license, and we may without any kind of censure bring any number of decrepid, imbecile, insane, scrofulous, and otherwise diseased children into the world; and yet from this ignorance and neglect arise half the ills that flesh is heir to. Darwin says: "For my part I conclude that of all the causes that have led to the differences in external appearance between the races of man, and to a certain extent between man and the lower animals, sexual relation has been by far the most efficient."* But Mr. Darwin throws very little additional light upon the laws that ought to regulate this selection between men and women, and I must confess to considerable disappointment, as I have been looking forward with great interest, expecting great revelations in his book on sex. We have not yet got much beyond the great

* "Descent of Man," vol. 2, p. 384.

principle of all breeders who have greatly improved our domestic animals, viz., that "like breeds like," although it is by the variations from this rule that the greatest improvements are effected. "Colour, form, size, texture of hair or wool, proportions of various parts, strength or weakness of constitution, tendency to fatten or to remain lean, to give much or little milk, speed, strength, temper, intelligence, special instincts; there is not one of these characters whose transmission is not an every-day occurrence within the experience of cattle breeders, stock-farmers, horse dealers, and dog and poultry fanciers." * No doubt in man, as in animals, we may improve a property by cultivation, and transmit it in its improved form to posterity; and this power may be greatly increased by a judicious sexual selection. Mr. Darwin tells us that "in the case of corporeal structures it is the selection of the slightly better-endowed and the elimination of the slightly less well-endowed individuals, and not the preservation of strongly marked and rare anomalies, that leads to the advancement of a species." † He says it is surprising how soon a want of care, or care wrongly directed, leads to the degeneracy of a domestic race; but excepting in the case of man himself, hardly any one is so ignorant as to allow his worst animals to breed. ‡ Much depends upon his choice of a wife, but a man's physical and mental condition also before marriage, and a woman's health and whole mental state during the period of gestation affect the coming child. Twins are often alike bodily and mentally, and yet differ widely from other members of the same family, which shows how much depends upon the mother and the circumstances in which she is placed. How little is at present known of, and how much less even do we care for, the natural laws by which beauty, health, and intellect result from certain unions,

* "Lay Sermons," p. 270. Huxley.
† "Descent of Man," p. 172. ‡ Ibid., p. 168.

and deformity, disease, and insanity from others.* We ought to know at what age it is best to marry; what physical and mental temperaments should be brought together, and what mode of treatment and feeding is best during the period of gestation. But of these things at present we have little or no record. Darwin, however, tells us that "children born of mothers during the prime of life are heavier and larger, and therefore probably more vigorous, than those born at other periods."† There can be no doubt that the military regulations, and the laws enforced with respect to marriage in many of the German States, have very much improved the

* "Our aristocracy, by exclusive intermarriages among ancient families, proceed blindly to breed in contempt of deformities, of feeble intellect, or of hereditary madness, under the instigation of pride, or the love of wealth, until their race becomes extinct; while another portentous curse, that of unwholesome factories, threatens to deteriorate the once brave manhood of England. I believe that, among mankind, as well as domesticated animals, there are physical and moral influences which may be regulated so as to improve or predispose both the corporeal and moral aptitudes; and certainly the most obvious cause is that of selecting the fit progenitors of both sexes."—(Sir A. Carlisle, in a letter to Mr. Alex. Walker "On Intermarriage.")

The Right Hon. B. Disraeli, in "Lothair," that rather clever satire of his on the upper classes, says: "It is the first duty of a state to attend to the frame and health of the subject. The Spartans understood this. They permitted no marriage the probable consequences of which might be a feeble progeny; they even took measures to secure a vigorous one. The Romans doomed the deformed to immediate destruction. The union of the races concerns the welfare of the commonwealth much too nearly to be entrusted to individual arrangement. The fate of a nation will ultimately depend upon the strength and health of the population. Both France and England should look to this; they have cause. As to our mighty engines of war in the hands of a puny race, it will be the old story of the lower empire and the Greek fire. Laws should be passed to secure all this, and some day they will be. But nothing can be done till the Aryan races are extricated from Semitism."—(Vol. 1, p. 260.)

† "Descent of Man," vol. 1. p. 174.

race during the last half century. "It is considered also as established, that residence in the Western States of America during the years of growth, tends to produce an increase of stature." The only case of "sexual selection" on record, except a few experiments lately made in America, is the well-known case of the Prussian Grenadiers, and this certainly produced a fine race of men physically, that is, judged by their inches, but they were proportionally stupid, growth of brain not having kept pace with growth of body. Where the vital force is used in growth of body, the thinking power is often small, and tall men have proportionally small brains. Mr. Galton regrets that he is unable to solve the simple question whether, and how far, men and women who are prodigies of genius are infertile. I have, however, he says, shown that men of eminence are by no means so. Eminence depends on position, circumstances, a large brain, perseverance, and power of concentration; but genius upon activity of the brain, and a general predominance of the nervous system. This absorbs the vital power, and especially in women prevents the due development of the generative system, and Mr. Galton will find that few women of any note have children.

We know so little that we do not even at present know what determines sex. The only thing that seems to have been established is that illegitimate births yield a smaller proportion of boys than girls.* In India the proportion of sexes is just the reverse to that which prevails in Europe. Omitting decimals, the males are to the females as 54 to 45; and in the North-West Provinces as 53 to 46; and in the rest of India as 51 to 48. In England, on the contrary, the ratio is 48 males to 51 females; and in Belgium the sexes are almost equally balanced.

What is wanted is a great collection of facts on this subject, that we may generalise the laws which control them.

* "Europe Politique et Social," by Maurice Block.

There has been no application at present of Darwin's very valuable labours to human nature. The Phrenologists are the only people who professedly have paid any attention to this subject. They have studied temperaments, muscular, vital, nervous, and their relation to character and to desirable sexual selections; but their knowledge is at present very limited. When more is known, however, there is little fear but that sensible people will attend to that upon which their own happiness and the welfare of the children so much depends. Perhaps when the world is full, and people live longer, we shall breed only from our Queen Bees, and put to death, as the bees do, at least for such purposes, all but our finest males.

Something is owing to Mr. Galton. He has demonstrated that both the physical and mental qualities are transmissible, by hereditary descent, in man as in other animals. He fears that the race is deteriorating under the demands that civilisation makes upon us, rather than advancing. He says, "the needs of civilisation, communication, and culture, call for more brains and mental stamina than the average of our race possess. We are in crying want of a greater fund of ability in all stations of life, for neither the classes of statesmen, philosophers, artisans, nor labourers are up to the modern complexity of their several professions. An extended civilisation like ours comprises more interests than the ordinary statesmen or philosophers of our present race are capable of dealing with, and it exacts more intelligent work than our ordinary artisans and labourers are capable of performing. Our race is overweighted, and appears likely to be drudged into degeneracy by demands that exceed its powers." Men are overworked, and require an artificial stimulant before the jaded nervous system is capable of enjoyment: men are over-excited, and require a sedative to quiet their too active brain, and both the disease and its palliative, alcohol and tobacco, are working infinite mischief. Over-toil, pain, sorrow, alcohol and tobacco, lower the health and decrease the size of the brain generally.

Under all these disadvantages, however, the child being born, we must proceed to develop, as far as we are able, all its physical and mental powers in harmonious proportion, depressing those that are too strong, and cultivating those that are weak. Our economical system aims more at the creation of wealth than the making of men. No doubt we might lose much in productive power by a general liability to military service, as is customary in Germany; but what we lost in wealth, we might gain in health and happiness. We turn our men into tailors and men-milliners, and keep them to occupations behind the counter, fit only for women; but it would tend greatly to the development of both their physical and mental powers if they were trained to handle a sword and rifle, instead of a needle and yard measure. Volunteering is only playing at soldiers. The attention given in this country to mere money-making is too exclusive, and a term of military service might well supplement the ordinary education, and give to Anglo-Saxon liberty what it most wants—a power of obedience, order, and organisation. It is true that our system of hiring people to fight for us absorbs a large and restless class that is little fit for any other occupation, and which might endanger the interests of the community by being let loose upon it; and neither is it likely that so large an interference with our money-making habits will be tolerated willingly for a moment; but still, should we ever be forced into it by self-defence, we may find some compensation in the physical training, and in the prevention of those early marriages which, with steam factories, are doing so much for the deterioration of the race in large towns.

All moral advance must be based upon improved physical conditions. What is called education exercises only the intellectual faculties, but it is health, prosperity, and happiness that are required to increase the strength of such feelings as make us wish to do justice and to love mercy. When a man is happy himself he wishes to make others so, and we are too apt to mete to others such measure as is given to us. More, however, after all, depends upon breeding than on culture, and it is as difficult, by taking thought, to add an inch to the mental as to the bodily stature. At present we systematically breed a race of paupers and criminals, for which the only palliative is a National System of compulsory education and training, and allowing labour to find its level, as water does, or as the price of other commodities, by affording every possible facility for its flowing freely from where it is in excess to where it is wanted. Our old system might be much mended, and last till we are better prepared for a new one, if all received a good technical education, and all could get easily to where their work was wanted, whether in this country or in the colonies.

Our happiness depends upon the exercise of all our powers, bodily and mental; and our Social System, as it at present exists, is by no means favourable to the division of our time and energies this varied exercise requires. Doubtless we have made some progress, but it is more in seeming than in reality. The *Tagespresse*, of Vienna, gives a curious extract from the Court regulations of the Hofburg for the year 1624, on the etiquette to be observed by officers when invited to the royal table. The regulation begins by stating that usually officers behave under such circumstances "with great politeness and good breeding, like true and worthy cavaliers;" but that the Emperor thinks it necessary to issue the following directions for the use of inexperienced cadets:—1. Officers should come to the palace handsomely dressed, "and not enter the room in a half-drunken state." 2. When they are at table "they should not rock about on their chairs, nor sit back and stretch out their legs." 3. They should not "drink after each mouthful, as by so doing they will very soon get drunk; nor drink more than half a glass at a time; and before drinking they should wipe their lips and moustachios." 4. They should not put their hands in the dishes nor throw bones under the table. 5. They should not "lick their fingers nor spit on the plate; nor wipe their noses with their napkins; nor drink so brutally as to fall off their chairs."

About the same time, or a little earlier, we have a Proclamation by the Lord Mayor of London, in which he set forth "that in divers places the players (Shakespeare and Company) do use to recite their plays to the great hurt and distraction of the game of bear-baiting, and such like pastimes, which are maintained for her Majesty's pleasure"; and a contemporary writer gives us a description of how our aristocracy at that time managed to get through their time in the country. "There are also of those," says Vincent, the defender of rural usages, "that can shoot in long-bows, cross-bows, or hand-gun: yea, there wanteth not some that are both so wise, and of so good audacity, as they can, and do (for lack of better company) entertain their master with table-talk, be it his pleasure to speak either of hawks or hounds, fishing or fowling, sowing or grafting, ditching or hedging, the dearth or cheapness of grain, or any such matters whereof gentlemen commonly speak in the country, be it either of pleasure or profit, these good fellows know somewhat in all." The brightest of them knew the rules of every game at dice and cards, could act in interludes, drink deeply for hours at a sitting without getting drunk, sing songs, make jests more pungent than those of professional fools, and on winter evenings, for the entertainment of august company, "read in divers pleasant books and good authors: as 'Sir Guy of Warwick,' 'The Four Sons of Aymon,' 'The Ship of Fools,' 'The Budget of Demands,' 'The Hundred Merry Tales,' 'The Book of Riddles,' and many other excellent writers both witty and pleasant." Yet further, a model serving-gentleman could in his master's absence play the part of host to visitors. In pre-Elizabethan times, we are told, these superior serving-men were always of gentle lineage.

The picture of the last century, also, is not very bright if we may trust Mr. Forsyth. He says: "As regards, however, the social aspect of the age, and the general tone of thought, it is, I think, impossible to deny that the bygone century is not an attractive period. There was little of the earnestness of life and quick invention and active benevolence which are the characteristics of our own age. The questions that have stirred the hearts of the present generation then slumbered in the womb of time. Reform, Free Trade, Education, and Sanitary Laws occupied no part of the thoughts of statesmen, and excited no interest in the people. The miracles of change which have been wrought by Steam, Electricity, Chloroform, Photography, and Breechloading Artillery, revolutionising Mechanics and Science, and Medicine, and Art and War, were not even suspected as possible."*

Our present advance, however, is not to be measured by these discoveries. It is not the physical but the moral advance with which we have to do. The ideal of our Upper Class tends to luxury and effeminancy; the suppression of all feeling and all mental power whatever; a noiseless anticipation of all personal wants by servants; enormous expenditure and display; and an unfailing adherence to a scale of social supremacy having little or no relation to goodness or talent. The country to this class, *without sport*, is still unquestionably a severe trial—the sport consisting, not in sympathy with the enjoyment around, but in killing and being their own butchers.†

* "The Novels and Novelists of the Eighteenth Century in Illustration of the Manners and Morals of the Age." By William Forsyth, M.A., Q.C. (London: John Murray, 1871.)

† A writer on "Social Slavery," in the *Cornhill Magazine*, tells us: "We stand at an evening party like skeletons in patent-leather boots,

Our Middle Class pass all their time in making money, that they may furnish a poor imitation of this. There is all the folly without the refinement.

and moralise upon the humbug and hollowness of the festivity that is going on around us. If all the crowd who are treading upon each other's toes could for one moment be induced to speak the truth what strange revelations would be made in a few minutes! What a quantity of petty jealousies and of mean ambitions would come to light! How small would be the results in real happiness to anybody concerned! How frankly we should confess that we were all boring each other to the very verge of despair, and that we would infinitely rather be at home, if it were not for some vague and unaccountable impression that it is the proper thing to bore and to be bored. Add all the toad-eating and the match-making and the place-hunting which, as novelists inform us, is taking place in every direction, and we shall be half inclined to believe that society is nothing but an organised system of hypocrisy, whose rules are made by clever rogues in order to provide a decent veil for their own trickery and selfishness. If people avowed their real motives the world would be too ugly a place to inhabit, and therefore an external decency is strictly enforced to cover its real deformity."

The *Saturday Review* remarks that sensational dramas, burlesques, and obscene dances are popular, not because Shakespeare and Sheridan are not played, but because Shakespeare and Sheridan are unintelligible and slow to the generality. The modern stage has not corrupted morals so much as our morals have corrupted the modern stage. Has it really come to this, that the middle-class, and indeed much of the higher, is incapable among us of more serious and ennobling amusements? Is it that the sense of better aims, purer enjoyments, and intellectual pursuits is dying out among us? There are ominous signs that patriotism is old-fashioned and out of date. A cynical contempt of a noble life, of higher literature, and of enthusiasm in any solid convictions is exhibited in the repose, as it is called, and indifference of manners and speech, which belong to high society. The highest culture among us is represented by a polite and superb contempt for any reality and heartiness; and among the middle classes the excessive devotion to wealth, and all that goes to creating personal wealth, absorbs the whole nature. Under these circumstances popular amusements and popular literature, burlesque, travesty, and comic journalism, the English stage and the English novel, are unpleasant phenomena. If the stage and the drama of the day reflect the morals of the day

The Lower and more numerous Class work so hard to furnish the means for the above most rational style of living (8,500 persons having an income of £126,000,000) that all their higher faculties get absorbed in their work—that is, in the vital and muscular force required for it.

The present tendency of our political institutions is towards disintegration, where every man, as an atom of society, shall have full power to do what he likes and to move about as he pleases; and this the Anglo-Saxon race in England, America, and elsewhere call freedom, although each man may still be the slave to his primary wants, and have to work hard for their gratification. When every one, however, has been so far freed it is to be hoped that Society will crystalise into a more perfect form. Real freedom is in obedience to law, under which we are *compelled* to act *voluntarily*. What is required to give real freedom to all, and to release the great body of the people from their too great and incessant toil, is organisation, under which a portion of our liberty would have to be given to the community that we might insure the larger part. The enormous saving of labour by organisation at the present time is well illustrated by the Post Office, where a letter is put in a hole in a wall, or pillar-post, and the answer to it, from any part of the kingdom, is dropped into a hole in your door next morning perhaps, one penny being sufficient to pay for the transit of this letter, and to leave a handsome profit on the transaction. Another illustration is where a Newspaper, which perhaps cost the proprietor from £20 to £100, is left at any man's door for a

we may almost begin to think that—at least as regards certain classes, the classes who want cheap amusements in great cities—we are on the verge, if not in the act of national decline. The ancient mimes of the dark days of the Triumvirs were almost identical with our farces and burlesques. The loose harlequinade and the looser dances of the Roman stage coincided with the days of the death struggles of the better Roman life. *Absit omen!*

penny. In fact, there is no want of this kind of organisation, by which, as Dr. Neil Arnott says, "each individual of the civilised millions dwelling on the earth may have nearly the same enjoyments as if he were the single lord of all";* but unfortunately these advantages are not at present given to each individual, but are confined to a favoured few—to men of "small fortune." Under better arrangements they might be extended to all. Society at present is a conglomeration of opposing interests. By the aid of machinery

* Dr. Neil Arnott, drawing a picture of what civilisation has done for us, says: "Every one feels that he is a member of one vast civilised society, which covers the face of the earth; and no part of the earth is indifferent to him. In England, for instance, a man of small fortune may cast his looks around him, and say with truth and exultation, 'I am lodged in a house that affords me conveniences and comforts which, some centuries ago, even a king could not command. Ships are crossing the seas in every direction, to bring me what is useful to me from all parts of the earth. In China, men are gathering the tea-leaf for me; in America, they are planting cotton for me; in the West Indies, they are preparing my sugar and my coffee; in Italy they are feeding silkworms for me; in Saxony, they are shearing the sheep to make me clothing; at home, powerful steam-engines are spinning and weaving for me, and making cutlery for me, and pumping the mines, that minerals useful to me may be procured. Although my patrimony was small, I have post-horses running day and night, on all the roads, to carry my correspondence; I have roads, and canals, and bridges, to bear the coal for my winter fire; nay, I have protecting fleets and armies around my happy country, to secure my enjoyments and repose. Then I have editors and printers, who daily send me an account of what is going on throughout the world among all these people who serve me. And in a corner of my house I have *Books!* the miracle of all my possessions, more wonderful than the wishing-cap of the Arabian Tales, for they transport me instantly, not only to all places, but to all times. By my books I can conjure up before me into vivid existence all the great and good men of antiquity; and for my individual satisfaction I can make them act over again the most renowned of their exploits: the orators declaim for me; the historians recite; the poets sing; and from the equator to the pole, or from the beginning of time until now, by my books, I can be where I please.'"

more work is now done in the cotton trade alone than formerly could have been done by the whole population of Great Britain. By such means we have become rich, but we have not yet learnt how to divide things fairly, or to make a right use of riches when acquired. The present system deprives the working man of the full benefit he might and ought to derive from such civilisation as we have. What is wanted is to carry the organisation further: to unite Capital and Labour, so that they should both work under one proprietorship. This would give to each individual the small fortune required, and the large surplus would not be wasted as it is now.

The small fortune that all might have under better arrangements would be quite sufficient to bring all our faculties into full activity. A large fortune is rather an impediment than an aid to the greatest happiness. Increase of luxuries does not mean increase of happiness, but decrease. It is not in possession, but in acquisition, that happiness is found—not in gratified wants, but in gratifying them. Fatigue is often only the too great strain on particular faculties; and rest is best found, not in inactivity, but in change of occupation, by bringing other faculties into use. A person actively engaged in the path of duty all day long is probably nearly as happy as he can be, while one who has nothing to do is nearly as miserable. "We are made up of activities nine parts, and passivities one—being capable of only one part in pleasure to nine parts in duty; and unless we prey upon something external, internal cravings will prey upon us. In other words, the satisfaction of active and benevolent exertions is almost inexhaustible; whereas pleasures cloy, and by repetition souring turn to pain. Labour is the doom of all. You may avoid the manual, but you increase the mental. You may avoid the burden of the shoulders, but you increase the burden of the heart and spirits." *

* "Elkinton Rectory," p. 415.

The question is, what are the wants that will bring all a man's faculties into daily activity? A man wants food, lodging, and clothing; a wife, children, and friends; a daily occupation, by which, with either head or hand, he produces something more than he consumes, or otherwise does his share of the world's work; and these should bring into activity all the energy of his nature, his love of overcoming difficulties, his policy, his planning, contriving, and constructing, and his love of acquisition so as to create some saving or surplus. A man has also pleasure in perseverance—in firmness; and he also requires a position in which he shall acquire, by truthfulness, honesty, and fair dealing, the approbation of his fellows and his own self-respect.

All this will bring one-half of a man's nature into full activity, but it is the lower half—the part he has in common with other animals. He requires some occupation that shall give him a strong interest in humanity—strong, if not as strong as he has in wife, children, and friends: an unselfish interest, as it is called—a hobby: an idea, as the French call it—an enthusiasm of humanity—something that will exercise his respect and veneration, and call out the faith, hope, and poetry of his nature. Above all, there must be the daily, serious, systematic cultivation of the intellect. To the well-cultivated intellect a single hour of the unveiling of nature by Science, of recreation in Art, or Music, or Poetry, is worth all the rest of the day. "Refined taste may be an equivalent for half an income, and a sense of what is beautiful in God's world may make a poor man

'Passing rich on forty pounds a year.'"*

"To have always," says Disraeli, "some secret, darling idea to which we can have recourse amid the noise and nonsense of the world, and which never fails to touch us in the most

* Rev. F. W. Robertson's "Lectures and Addresses."

exquisite manner, is an act of happiness that fortune cannot deprive us of."

Frederick Robertson says most truly that "there is something worse than death: Cowardice is worse. And the decay of enthusiasm and manliness is worse. And it is worse than death, aye, worse than a hundred deaths, when a people has gravitated down into the creed that the 'wealth of nations' consists not in generous hearts—'fire in each breast and freedom on each brow'—in national virtues, and primitive simplicity, and heroic endurance, and preference of duty to life;—not in MEN, but in silk and cotton, and some-that they call 'capital.' Peace is blessed. Peace arising out of Charity. But peace, springing out of the calculations of selfishness, is not blessed. If the price to be paid for peace is this, that wealth accumulates and men decay, better far that every street in every town of our once noble country should run blood."*

But mankind generally at present do not get much beyond the activity of the lower half of their faculties. It is true that on Sundays they do their Religion, but this is too frequently little more than a selfish and prudent endeavour to "make the best of both worlds." They put off till a more convenient season the action of their higher powers, but that convenient season never comes, or if it does come, the higher powers have disappeared for want of cultivation. "Capacity for the nobler feelings is in most natures a very tender plant, easily killed, not only by hostile influences, but by mere want of sustenance." (J. S. Mill.)

Herbert Spencer defines life to be "The continuous adjustment of internal relations to external relations." These undoubtedly are the conditions of life, but more especially is this adjustment essential to the harmonious development of mental life. There is no fact perhaps more overlooked than

* Rev. F. W. Robertson's "Lectures and Addresses," p. 196.

this in the plans we lay out for our own well-being. There is a time for all things, but when that time is passed, those things are gone from us for ever. In the eager pursuit of certain objects in life, the end is often lost in the means. In the pursuit of wealth the use of that wealth is too often forgotten, till not the use, but the pursuit alone gives pleasure. The moral and æsthetic nature has been crippled, if the soul has not been sold altogether to the devil of this world—the god Mammon. Each of the seven ages of life has its joys allotted to it, which, if not taken at the time, are lost to us for ever. Some people never catch the proper time; they are always an age behind, and the true enjoyments of life are, consequently, never known to them. The continuous adjustment of external relations to internal relations is forgotten, and procrastination robs them of joys which have been put off until constitutional changes have rendered such joys no longer possible. It is this harmony of faculty and circumstance that is essential to happiness. Few know how much depends upon the nice adjustment of our pursuits to our varying bodily and mental capacities: it is useless to try to put old heads on young shoulders, or to do in old age what only youth fits us for. That "some place their bliss in action, some in ease," depends upon age. If we would not throw away happiness, and indeed make ourselves miserable, we must make careful note of these changes, and adapt our pursuits to our altered or changing capabilities, or we shall soon discover that "all is vanity." In youth the propensities are active—"we draw nutrition, propagate," and if no higher faculties are cultivated, "rot," at least mentally.

> Ah me! what wonder-working, occult science
> Can from the ashes in our hearts once more
> The rose of youth restore?
> What craft of alchemy can bid defiance
> To time and change, and for a single hour
> Renew this phantom-flower?

"O give me back!" I cried, the vanished splendours,
The breath of morn and the exultant strife,
When the swift stream of life
Bounds o'er the rocky channels, and surrenders
The pond, with all its lilies, for the leap
Into the unknown deep!

And the sea answered, with a lamentation
Like some old prophet wailing, and it said,
"Alas! thy youth is dead!
It breathes no more, its heart has no pulsation,
In the dark places with the dead of old
It lies for ever cold!"

—Longfellow.

But "though nothing can bring back the hour of sunshine in the grass, of glory in the flower," this sunshine and glory are more dependent upon cultivation than upon age, and ought to increase rather than decrease as we grow older. The Æsthetic feelings are the result of careful cultivation, and should be the solace of old age. The largest fortune we can lay by for our old age is a taste for good reading, the result of careful cultivation. To keep the company of all the great and good men that have lived is the best "society" we can have. The stores of literature are inexhaustible, and he who has acquired a taste for literature has never a minute that he cannot spend pleasurably. The proper love of advancing age is the love of truth, and the proper business of old age is its discovery. The wealth of such a love and such a pursuit is endless. The world is full of new books and new truths, and fresh store is laid at our feet every morning. We have only to put ourselves *en rapport* with all that is great, wise, and good in the world to form a part of it. A mere intellectual sensualism is not what I mean. Gobbling up a great number of books without order, or arrangement, or digestion, is of little use. We should read good books, pencil in hand, to note what we wish to return to, and retain and put down thoughts that may be suggested. We gain more by the latter often than the former. With a habit of good reading, a good novel, as an occasional relaxation, may, perhaps, give to others, as it does to me, the greatest possible enjoyment; and it is certainly the cheapest pleasure we can have, because if we do not indulge too freely there is no expense and no drawback. We must lay down a plan of life, and have strength of will enough to keep it; and to do this it is essential that we should know definitely what we would have, and if we attain to that we must be satisfied, and not be disappointed if we cannot also have other things that are incompatible with it. For instance, if truth is our object, we must learn to do without the "stupid staring and the loud huzzas" of the multitude. "The highest object in life," says Samuel Smiles, "we take to be, to form a manly character, and to work out the best development possible of body and spirit—of mind, conscience, heart, and soul, &c. That is, therefore, not the most successful life in which a man gets the most pleasure, the most money, the most power or place, honour, or fame, &c."* We must get out of the deep rut of conventionalism, and only move in it when it saves time instead of fritters it away, or we may have no solitude, and our leisure be at the mercy of any person who may choose to inflict his idleness or *ennui* upon us. Who cannot feel with Longfellow, his "Day of Sunshine" has not yet learned to Live!—

O gift of God! O perfect day:
Whereon shall no man work, but play;
Whereon it is enough for me,
Not to be *doing*, but to be!

* "Self Help," p. 253.

Through every fibre of my brain,
Through every nerve, through every vein,
I feel the electric thrill, the touch of
Life, that seems almost too much.

I hear the wind among the trees
Playing celestial symphonies;
I see the branches downward bent
Like keys of some great instrument.

And over me unrolls on high
The splendid scenery of the sky,
Where through a sapphire sea the sun
Sails like a golden galleon.

As we get older we live more in the ideal world than in the real. Present impressions are weak, memory fails, and we fall back upon our early life. The secret of our emotions never lies in the bare object, but in its subtle relations to our own past. No wonder the secret escapes the unsympathising observer, "who might as well put on his spectacles to discern odours."* How much, then, depends upon the preparation we have made—upon cultivation and association. "A youth of follies, an old age of cards," is the too common lot, when we ought to be reposing in "the soul's calm sunshine and the heartfelt joy."

In old age we also lose the elastic spring of youth and our disposition to wander. As the breeze over the trees, as the sunshine over the flowers, as the waves over the sea-anemone as he remains fixed to the rock, so the panoramic world passes over us, instead of our passing over the world; and, like the anemone, we spread out our feelers, absorbing all that is requisite for our mental assimilation.

And let us not forget that for the cultivation of the higher feelings, especially of the Æsthetic, Solitude is a necessity. It is good to be alone. What prayer is to the fervour of devotional feeling, to be alone with Nature is to the Æsthetic.

* George Eliot.

All appearance, whatever we see in the world, is but as a vesture of that which lies at the base of appearing—God; and there are moments when we give ourselves up to solitude and nature, in which the veil seems partially withdrawn, and everything around us takes a Divine aspect, and every lichen, moss, and flower, every bird and butterfly, the blue sky and the first peeping star, are received into our bosoms in a spirit of universal love,—when all that is individual and personal is felt to be a part of, if not absorbed into, the Infinite Whole. This is true Religious feeling. F. Robertson, who knew well this feeling, says: "I am alone, lonelier than ever,—*sympathised with by none, because I sympathise too much with all.* But the All sympathises with me—a sublime feeling of a Presence comes over me at times, which makes universal solitariness a trifle to talk about."*

* George Sand speaks under similar influences: "The moments in which, possessed and carried out of myself by the power of external things, I can abstract myself from the life of my species, are absolutely fortuitous, and it is not always in my power to make my soul pass into beings which are not myself. When this phenomenon produces itself spontaneously, I could not tell whether any particular circumstances, psychological or physiological, has prepared me for it. It certainly requires the absence of strong preoccupation; the least cause for solitude keeps away this kind of inward ecstasy, which is like an unforeseen and involuntary forgetfulness of my own vitality. No doubt this happens to everybody, but I should like to meet with some one who could say to me: 'This also happens to me in the same manner. There are hours when I escape from myself, when I live in a plant, when I feel myself grass, bird, tree-top, cloud, running water, horizon, colour, form and sensation changing, moving, indefinite; hours in which I run, I fly, I swim, I drink the dew, I expand in the sun, I sleep under the leaves, I soar with the larks, I crawl with the lizards, I shine in the stars and the glow-worms—in short, I live in all that is the medium of a development which resembles an extension of my being.' I have never met one who spoke thus, or have met him without knowing him. I could, however, have wished to meet him on condition that he was wiser than I, and could tell me whether these phenomena are the result of a bodily or mental condition? whether they arise from the instinct

We may plan our relations to our fellow-man; but our relation to God is our relation to the Universe, and that properly is the Religion of the Universe, the most important to us of all. We see that Worlds are all governed from a centre by what we call the attraction of gravitation, but we do not equally recognise that every atom is so also. Victor Cousin truly says, "En effet si Dieu n'est pas tout, il n'est rien." God is everything or nothing; and "the conception of the Universe is incomplete, if not comprehended as the entire and continuous work of the eternally-creating Spirit."* "Science is the history of the Divine operations, and the world with all its antiquity is yet every moment a new creation." But this is accepted by Theologians only in theory. It has never been, as it ought to have been, a practical proposition. The universe must be regarded as the "manifestation of some transcendent life, to which our separate individual life is related," and from which indeed it cannot exist apart. Man is nothing in himself: he is a part only of a larger system to which he must conform. He must work *in* this, or be forced *out of it* altogether. If he works smoothly he is happy; if he does not he is miserable. He is part of the elements around him, and can only work in conjunction with them, from which his power is derived. He borrows as much of this power—which is God's power—as is necessary to effect his purposes. To do this he must act in accordance with the laws by which this power is manifested. It is lent to him only on these conditions. If he accepts these conditions, he may succeed; if he does not, he must certainly fail. His true interest, then, is to learn

of universal life which physically resumes its rights over the individual, or if they have a higher parentage, an intellectual relation to the soul of the universe which reveals itself to the individual delivered at certain hours from the ties of personality. My opinion is that there is some of both the one and the other, and even that it cannot be otherwise."—(*Pall Mall Gazette*, Sept. 23, 1871.)

* Oersted.

what these conditions are, and the Revelation and its true interpretation are placed within his reach.* As we are connected by the force of gravitation to all around, so our other forces, odylic, electric, vital, mental—although this is not so evident to us—are equally connected with all around, and we must put ourselves *en rapport*, and act harmoniously with them. "The universal soul is the alone creator of the useful and the beautiful; therefore to make anything useful or beautiful, the individual must be submitted to the universal mind."† We must follow Nature, who in the infinite ages has tried all ways, and transmitted only the best to us. "Dolland formed his achromatic telescope on the model of the human eye," and man is never left without precedent if he will seek for it; and if we will imitate our Father in Heaven we must follow Nature.

* "Everything good in man leans on what is higher. This rule holds in small as in great. Thus all our strength and success in the work of our hands depend upon our borrowing the aid of the elements. You have seen a carpenter on a ladder with a broad axe chopping upward chips from a beam. How awkward! at what disadvantage he works! But see him on the ground, dressing his timber under him. Now, not his feeble muscles, but the force of gravity brings down his axe; that is to say, the planet itself splits his stick. The farmer had much ill-temper, laziness, and shirking to endure from his hand-sawyers, until one day he bethought him to put his saw-mill on the edge of a waterfall; and the river never tires of turning his wheel: the river is good-natured and never hints an objection. * *

"Now that is the wisdom of a man, in every instance of his labour, to hitch his waggon to a star, and see his chore done by the gods themselves. That is the way we are strong, by borrowing the might of the elements. The forces of steam, gravity, galvanism, light, magnets, wind, fire, serve us day by day, and cost us nothing. * *

"All our arts aim to win this vantage. We cannot bring the heavenly powers to us; but, if we will only choose our jobs in directions in which they travel, they will undertake them with the greatest pleasure. It is a peremptory rule with them, that *they never go out of their road*: they never swerve from their preordained paths,—neither the sun, nor the moon, nor a bubble of air, nor a mote of dust."—(Emerson, "Society and Solitude. Civilisation," pp. 22, 24.

† Emerson.

The advance of the race must be measured by the increased and increasing size of the nervous system, and civilisation by the increased facilities for bringing that system into legitimate action, and for keeping this "harp of a thousand strings" in tune. "The world is an organic whole under natural laws," and civilisation*—about the true meaning of which there is still so much dispute—must be measured by the extent to which this whole can be brought into harmonious action. Man must continue to advance: what may be his ultimate destiny it is impossible to say.†

As nine months go to the shaping an infant ripe for his birth,
So many a million of ages have gone to the making of man:
He now is first, but is he the last? is he not too base?

—Tennyson.

* It was suggested in the Statistical Section of the British Association that the degree of civilisation at present existing in any country is proportioned to the quantity of soap there consumed, so that, as it was observed by the noble lord presiding, the question "How are you off for soap?" is not perhaps so irrelevant and inappropriate as it may sometimes seem.

† Some of our philosophers consider man at present only in the grub state. Thus Mr. J. W. Jackson, F.A.S., tells us (*Anthropological Review*, 1867,) that "every *mode* of being has its own *sphere*," and that Man "is the beginning of a new *order*, the bipedal and aerial type of the mammal, but of this he is obviously an immature and merely germal specimen," and that as the grub becomes a butterfly and the reptile a bird, so will man, in the process of development, drop his too predominant vascular arrangements and alimentary functions, and fly too." "And," says Mr. Jackson, "what a stupendous vista of progress and possibility is thus opened to our wondering gaze! Man but the unfledged *beginning* of a new order of being, the callow nestling of the future eagle of the skies; the precursor, and in a sense the progenitor, of earth's manifold types of intelligent being. * * * Thus contemplated, then, we also see that existing Man is not the Divine idea of Humanity in its final form, but only that idea in the process of realisation."

I must leave Mr. Darwin, Mr. Wallace, and Mr. Jackson to *settle* this between them, having no opinion of my own. No doubt much may yet be expected from Natural and Sexual Selection. Electro-biology, Mesmerism, the so-called Spiritual Phenomena, &c., point to

"There are diversities of operations, but the same God worketh all in all." Now, what are these "diversities of operations," or, as Science calls them, Modes of Motion? The best arrangement and classification of things or objects, and of the changes going on among them, have been given to us by Dr. Neil Arnott:—

### A GENERAL OUTLINE MAP OR TABLE OF MAN'S KNOWLEDGE OF NATURE.

---

FIRST DIVISION.—NATURAL THINGS OR OBJECTS.

| | | |
|---|---|---|
| THE THREE KINGDOMS of Nature. | ANIMAL . . | Man and lower races of Animals. |
| | VEGETABLE | Trees, Plants, Flowers, and Fruits. |
| | MINERAL . | Stones, Earths, Metals, and other lifeless things. |

SECOND DIVISION.—NATURAL CHANGES GOING ON AMONG OBJECTS.

| | |
|---|---|
| The four orders of changes or phenomena among THINGS, knowledge of which is called SCIENCE. | PHYSICAL,—Natural Philosophy. |
| | CHEMICAL,—ElementarySubstances. |
| | VITAL,—Life and Health. |
| | MENTAL,—Intellect and Happiness. |

THIRD DIVISION.—ARTIFICIAL CHANGES PRODUCED BY HUMAN INGENUITY.

| | |
|---|---|
| The ARTS by which man facilitates and stores knowledge. | LANGUAGE with Alphabet. |
| | COUNTING and MEASURING |

| | |
|---|---|
| Industrial Arts . . . . . | Agriculture. |
| | Rearing tame Flocks and Herds. |
| | ENGINEERING, as spinning, weaving, &c., &c., &c. |

a great advance in the direction Mr. Jackson has indicated; showing that the nervous force, that resumes its conciousness in the brain, need not necessarily come through the stomach, but that this result of cerebration might be derived by way of inspiration *directly* from without. In such case we might, as Mr. Jackson observes, "drop our too predominant vascular arrangements and alimentary functions," and our low habits of eating and drinking, and, coming events

Here is a clear chart of all the knowledge we can attain. It will all come under these heads,* and five good books might contain it all. The sciences are so mutually related that if a person has not mastered five such books he is not fully qualified to give an opinion on any subject, as he will necessarily be ignorant of the laws and principles upon which a competent opinion should be based. We have here an account of all the motions going on around us to which our knowledge yet extends, and our existence and well-being will be secured in proportion as our own motions harmonise with these. Our own Consciousness is alone *directly* known to us, but these motions are known to us by inference, and the cause of these motions are equally known to us by inference. For as "where there is motion something is moved," so where there is power there is something of which it is the power, and this is God. Behind this phantasmagoria of a

having cast their shadows before, we should probably resemble the cherubs we have so long been accustomed to see floating over our altar-pieces, all head and wings; and as in the process of transformation the greedy silkworm spins his entrails into silk, alone worthy to cover the most divine form in nature, so may man spin his into etherial cat-gut for some superior kind of instrument to harmonise with the music of the spheres; and as, also, "the silkworm spins its cocoon by a decomposition or retrograde chemical action, which imparts to the remaining blood a higher vital status fitted for the more elevated grade on which it is about to enter, and the vital condition of the insect is elevated to a correspondence with the demands of the higher organisation it is destined to assume," ("Life in Nature," J Hinton, pp. 73, 241); so may man's brain be raised to a higher power and temperament, and Inspiration and Clairvoyance be its normal condition. Not being able to sit down, not having *de quoi*, may necessitate some little alteration in our present social customs. But as this desirable change is not expected to take place *immediately*, every one may safely be left to form his own views, which must undoubtedly at present contain a considerable amount of speculation.

* See Dr. Arnott's fuller exposition of this Table in the Introduction to his "Elements of Physics;" the most useful single acquisition perhaps that the mind can make.

world—this ever-changing Kaleidoscope—we find a God producing a *real* world of enjoyment. "This fair universe, in the meanest province thereof, is in very deed the star-domed City of God; through every star, through every grass-blade, and most through every living soul, the glory of a present God still beams. But Nature, which is the Time-vesture of God, and reveals Him to the wise, hides Him from the foolish."* We have "Creation" going on all around us, for every fresh effect is a new creation. "The breezy, incense-breathing morn" is born afresh every day, so is young love; the "primrose by the river's brim" and the lark are immortal, and yet are new creatures every spring, so are our affections and emotions; if we all die, we are all young, and all that is good and beautiful and profitable, as the days and seasons change, are laid at our feet for our acceptance. The breeze is as fresh as that which passed over Eden, and the sun-sets as glorious as those which accompanied "still evening" to our first parents. It is we are changed, not they: our power of appreciation has increased and must continue to increase with each succeeding generation—with each change of body. We are not obliged to rest in a problematical history of Creation 6,000 years ago. The supposed difference between secular and religious has no real existence. Whatever other Revelation we may have, we have God ever present revealing Himself in every cause and effect, in every evolution, in every manifestation of power, and in every change or mode of motion that attends it. We must give a sacred and religious character to His mode of working, and read a chapter in the Bible of His doings every day, and thus every day turn some new leaf in Science. Here we may see God face to face, for His laws are His attributes, and we must break through the anthropomorphism of our childhood and learn to recognise Him. It is not the "untutored

* Carlyle. "Sartor Resartus," p. 274.

mind" that "sees God in clouds and hears Him in the wind," but the most cultivated and the most religious.

The "One and all" requires the resignation of the individual and personal—of all that is selfish—to the Infinite Whole. Man loses himself in the Divine, and he must act harmoniously with it. "In the Church of the latter days," says St. Simon, "man is to feel and realise the divinity of his whole nature, material as well as spiritual." This is the Key-stone of the New Religion, of which Spinoza, as he has done most to embody the idea, and Atheist as he has been called, must be accepted as the High Priest. "Naturæ convenienter vivere" is our motto, and true humility is the recognition of the greatness of the whole.

The practice of this Religion is very simple. "We must learn what is true that we may do what is right." Right must take precedence of everything else, and that only is right which is in accordance with Natural laws—that is, with God's mode of working; for that most directly leads to the greatest good. It is this that makes it right. The Calvinists have put "Sovereign Will" as the measure of right, but that would be no rule of right for us unless it tended to the general good; right must always govern will, not will right, even in the Highest. As right is obedience to Law upon which the general happiness depends, and as nothing can continue long to exist that is not in harmony with such Law, right and might are the same. Of course this practice very much extends the boundary of ordinary morality and duty, for right claims equal obedience to all law—physical, moral, and intellectual. We cannot break any law, either voluntarily or involuntarily—from free will or necessity—without being made to suffer what has been called punishment; but the suffering is to teach us what is right and to enforce obedience. It is of no use praying to be let off: God makes no exception. He is no respecter of persons; He never forgives; and as punishment is for our good, it would be an injury to us if He did.

From its earliest days a child should require no other reason for its conduct than because it is right. It will often have to trust to its parents to say what *is* right, and the same trust and faith and reverence should be transferred in after life to Natural Law—that is, to God. "I hold," says Prof. Huxley, in his letter to the Rev. W. H. Freemantle on the duties of the London Education Board, "that any system of education which attempts to deal only with the intellectual side of a child's nature, and leaves the rest untouched, will prove a delusion and a snare, just as likely to produce a crop of unusually astute scoundrels as anything else. In my belief, unless a child be taught not only that doing what is right is wise (which is morality), but also that the right is above all things beautiful and lovable (which, as I understand the term, is religion), education will come to very little." With this I think we all ought to agree. As the business of infancy is to connect the bodily movements with the mind, so the main educational business of later life is to connect all our movements in the same way with our sense of right. Justice is the great key-stone of the moral arch, and should take precedence of all other virtues—that is, it is more "right" than all others. We may be often willing to put up with less than justice, and to take less than actually belongs to us, but it should be always under protest. It is this system of exact equivalents upon which the moral world moves, and is as necessary as the balance of action and reaction in the physical world. It is justice, or what is *due to* others, that prevents our passions and individual interests from trespassing upon their equal rights. Love is the sunshine of life, and hard though it may be, we can live without it; but we can no more live without justice than without the sun itself. It is this system of equivalents upon which moral chemistry is based. We must not only be just before we are generous, but just before all; for anything short of the claims of justice is so much taken from others to which they are entitled. We may please ourselves whether we will give, but we have clearly no right to take away that which belongs to another. What a change it would immediately make in the world if no man by speech or action robbed another of what was due to him,—of neither his time, his fair fame, his labour, or his goods.

I have been the more solicitous thus to insist upon the claims of Justice, because a spurious Benevolence during the Christian era has been sapping the very foundations of morality. A charity that is not just, has been undermining self-reliance, self-dependence, and self-respect, and damaging the best interests of society. We are constantly placing ourselves between a man's actions and their natural consequences, and the effects have been such that it cannot be too often repeated that we must learn what is true, that we may do what is right.

THE END.

# INDEX.

www.ingramcontent.com/pod-product-compliance
Lightning Source LLC
Chambersburg PA
CBHW021359210326
41599CB00011B/938

* 9 7 8 3 7 4 1 1 5 5 6 0 4 *